The Fragmentation of Global Climate Governance

NEW HORIZONS IN ENVIRONMENTAL AND ENERGY LAW

Series Editors: Kurt Deketelaere, *Professor of Law, University of Leuven, Belgium and University of Dundee, Scotland* and Zen Makuch, *Reader in Law, Barrister, Imperial College, London, UK*

Environmental law – including the pressing considerations of energy law and climate change – is an increasingly important area of legal research and practice. Given the growing interdependence of global society and the significant steps being made towards environmental protection and energy efficiency, there are few people untouched by environmental and energy lawmaking processes.

At the same time, environmental and energy law is at a crossroads. The command and control methodology that evolved in the 1960s and 1970s for air, land and water protection may have reached the limit of its environmental protection achievements. New life needs to be injected into our environmental protection regimes – perhaps through the concept of sustainability in its environmental, economic and social forms. The same goes for energy policy and law, where liberalization, environmental protection and security of supply are at the centre of attention. This important series seeks to press forward the boundaries of environmental and energy law through innovative research into environmental and energy law, doctrine and case law. Adopting a wide interpretation of environmental and energy law, it includes contributions from both leading and emerging international scholars.

Titles in the series include:

International Law and Freshwater
The Multiple Challenges
Edited by Laurence Boisson de Chazournes, Christina Leb and Mara Tignino

Genetic Resources, Equity and International Law
Camena Guneratne

Energy for the 21st Century
Opportunities and Challenges for Liquefied Natural Gas (LNG)
Susan L. Sakmar

Emissions Trading Design
A Critical Overview
Stefan E. Weishaar

EU Environmental Legislation
Legal Perspectives on Regulatory Strategies
Edited by Marjan Peeters and Rosa Uylenburg

Polar Oceans Governance in an Era of Environmental Change
Edited by Tim Stephens and David L. VanderZwaag

The Fragmentation of Global Climate Governance
Consequences and Management of Regime Interactions
Harro van Asselt

The Fragmentation of Global Climate Governance

Consequences and Management of Regime Interactions

Harro van Asselt

Stockholm Environment Institute (SEI)

NEW HORIZONS IN ENVIRONMENTAL AND ENERGY LAW

Edward Elgar

Cheltenham, UK • Northampton, MA, USA

Published by
Edward Elgar Publishing Limited
The Lypiatts
15 Lansdown Road
Cheltenham
Glos GL50 2JA
UK

Edward Elgar Publishing, Inc.
William Pratt House
9 Dewey Court
Northampton
Massachusetts 01060
USA

A catalogue record for this book
is available from the British Library

Library of Congress Control Number: 2013958024

This book is available electronically in the ElgarOnline.com Law Subject Collection, E-ISBN 978 1 78254 498 2

MIX
Paper from
responsible sources
FSC
www.fsc.org
FSC® C013056

ISBN 978 1 78254 497 5

Typeset by Columns Design XML Ltd, Reading
Printed and bound in Great Britain by T.J. International Ltd, Padstow

For Rowena, for making the journey that is my life so much better.

Contents

Tables

Acknowledgements

Writing a book can be a solitary exercise, but I have been lucky enough to have worked in excellent research environments over the years. The initial research for this book was conducted at the Institute for Environmental Studies at the VU University Amsterdam. This was followed by a Marie Curie Fellowship (CLIMATEGOV – Contract no. 253090) with the Environmental Change Institute at the University of Oxford. The research was completed at the Stockholm Environment Institute in 2013. I am very grateful to the many friends and colleagues who have supported me throughout this endeavour.

I have had the privilege of testing ideas with the following people through discussions, feedback on written work or co-authorship: Dan Bodansky, Tom Brewer, Harriet Bulkeley, Heleen de Coninck, Thijs Etty, Liz Fisher, Nicolien van der Grijp, Aarti Gupta, Dave Huitema, Andrew Jordan, Norichika Kanie, Sylvia Karlsson-Vinkhuyzen, Kati Kulovesi, Andrew Long, Nele Matz-Lück, Jeff McGee, Michael Mehling, Ayşem Mert, Remi Moncel, Sebastian Oberthür, Philipp Pattberg, Jerneja Penca, Benjamin Simmons, Francesco Sindico, Johannes Stahl, Claire Stockwell, Olav Schram Stokke, Ludivine Tamiotti, Kyla Tienhaara, Antto Vihma and Christina Voigt.

I should give special credit to a few people in particular: Frank Biermann and Joyeeta Gupta, whose encouragement to complete this book has been indispensable; Fariborz Zelli, who through friendship and co-authorship has greatly inspired my thinking on regime interactions; and Constanze Haug, who has helped me through many difficult stages of the writing process.

Edward Elgar Publishing has been supportive since the very beginning, and I would like to particularly thank Matthew Pittman for his encouragement, understanding and, above all, patience.

Parts of this book builds on published work, which has been updated, adapted and largely rewritten.

Sections of Chapters 3 and 9 were published as: van Asselt, Harro (2012), 'Managing the Fragmentation of International Climate Law', in Erkki Hollo, Kati Kulovesi and Michael Mehling (eds.), *Climate Change and the Law*, Berlin: Springer, 329–357.

Sections of Chapter 5 are updated from: van Asselt, Harro (2011), 'Legal and Political Approaches in Interplay Management: Dealing with the Fragmentation of Global Climate Governance', in Sebastian Oberthür and Olav Schram Stokke (eds.), *Managing Institutional Complexity: Regime Interplay and Global Environmental Change*, Cambridge, MA: The MIT Press, 59–85.

Sections of Chapter 6 draw on: van Asselt, Harro (2007), 'From UN-ity to Diversity? The UNFCCC, the Asia–Pacific Partnership, and the Future of International Law on Climate Change', *Carbon and Climate Law Review*, 1(1), 17–28; van Asselt, Harro, Norichika Kanie and Masahiko Iguchi (2009), 'Japan's Position in International Climate Policy: Navigating between Kyoto and the APP', *International Environmental Agreements: Politics, Law and Economics*, 9(3), 319–336; and Karlsson-Vinkhuyzen, Sylvia and Harro van Asselt (2009), 'Introduction: Exploring and Explaining the Asia–Pacific Partnership on Clean Development and Climate', *International Environmental Agreements: Politics, Law and Economics*, 9(3), 195–211.

Chapter 7 is a shortened and updated version of: van Asselt, Harro (2012), 'Managing the Fragmentation of International Environmental Law: Forests at the Intersection of the Climate and Biodiversity Regimes', *New York University Journal of International Law and Politics*, 44(4), 1205–1278.

Sections of Chapter 8 have been published as: van Asselt, Harro and Thomas L. Brewer (2010), 'Addressing Competitiveness and Leakage Concerns in Climate Policy: An Analysis of Border Adjustment Measures in the US and the EU', *Energy Policy*, 38(1), 42–51; and Zelli, Fariborz and Harro van Asselt (2010), 'The Overlap between the UN Climate Regime and the World Trade Organization: Lessons for Post-2012 Climate Governance', in Frank Biermann, Philipp Pattberg and Fariborz Zelli (eds.), *Global Climate Governance Beyond 2012: Architecture, Agency and Adaptation*, Cambridge, UK: Cambridge University Press, 79–96.

Finally, sections of Chapter 10 draw on: Moncel, Remi and Harro van Asselt (2012), 'All Hands on Deck! Mobilizing Climate Change Action beyond the UNFCCC', *Review of European Community and International Environmental Law*, 21(3), 163–176; and Zelli, Fariborz and Harro van Asselt (2013), 'The Institutional Fragmentation of Global Environmental Governance – Causes, Consequences and Responses', *Global Environmental Politics*, 13(3), 1–13.

Abbreviations

APP	Asia–Pacific Partnership on Clean Development and Climate
Art.	Article
AWG-KP	Ad Hoc Working Group on further commitments for Annex I Parties under the Kyoto Protocol
AWG-LCA	Ad Hoc Working Group on Long-term Cooperative Action
BAM	border adjustment measure
CBD	Convention on Biological Diversity
CDM	Clean Development Mechanism
CO_2	carbon dioxide
COP	Conference of the Parties
COP/MOP	Conference of the Parties serving as Meeting of the Parties
CTE	Committee on Trade and Environment
EU	European Union
G8	Group of 8
G20	Group of 20
G77	Group of 77
GATS	General Agreement on Trade in Services
GATT	General Agreement on Tariffs and Trade
GDP	gross domestic product
ILC	International Law Commission
ILM	International Legal Materials
IPCC	Intergovernmental Panel on Climate Change
LULUCF	land use, land-use change and forestry
MEA	multilateral environmental agreement
OECD	Organisation for Economic Co-operation and Development
para.	paragraph

PIC	Policy and Implementation Committee
ppm	parts per million (volume)
PPMs	processes or production methods
REDD	reducing emissions from deforestation and forest degradation
R&D	research and development
SCM Agreement	Agreement on Subsidies and Countervailing Measures
TBT Agreement	Agreement on Technical Barriers to Trade
TRIMS Agreement	Agreement on Trade-Related Investment Measures
TRIPS Agreement	Agreement on Trade-Related Aspects of Intellectual Property Rights
UN	United Nations
UNCLOS	United Nations Convention on the Law of the Sea
UNEP	United Nations Environment Programme
UNFCCC	United Nations Framework Convention on Climate Change
UNGA	United Nations General Assembly
UN-REDD	United Nations Collaborative Programme on Reducing Emissions from Deforestation and Forest Degradation in Developing Countries
UNTS	United Nations Treaty Series
US	United States
VCLT	Vienna Convention on the Law of Treaties
WTO	World Trade Organization

Table of cases

Table of treaties and decisions

TREATIES

DECISIONS BY THE CBD COP

DECISIONS BY THE KYOTO PROTOCOL COP/MOP

DECISIONS BY THE UNFCCC COP

PART I

Setting the scene

1. Introduction

1. COPING WITH COMPLEXITY

Comprehension of the interconnectedness of human behaviour and the natural environment poses a crucial challenge in our efforts to govern global environmental problems such as climate change, biodiversity loss, deforestation and ozone depletion. Although the causes and consequences of, and responses to, these problems are inevitably linked with each other and with other issue areas, laws and policies at various levels of governance have largely developed in isolation. This has led to a mismatch between the simplified way in which humans govern their relationship with the environment and the complexity of biogeophysical systems. In an era of globalization, governing human behaviour not only has to account for the world's ecological interconnectedness but also for social, economic and political interdependence.[1] In essence, this book is about coping with the resulting socio-ecological complexity in global environmental governance.

The challenge of dealing with socio-ecological complexity is especially profound in the governance of global climate change. Climate change can be considered a 'wicked problem' *par excellence*.[2] In part, this is because there is no clear definition of what the 'problem' is. Is it, for instance, about reducing greenhouse gas emissions, about phasing out fossil fuel use, or is the problem bigger than that, for example the continued insistence on economic growth?[3] Similarly, there is no simple 'solution' for the climate change problem, as solving the problem of climate change will depend on how one defines the problem in the first place. Any solution will thus be subject to value judgements. Moreover, solutions to wicked problems like climate change are likely to have a ripple effect, potentially causing new problems in the process. Once a

[1] O.R. Young *et al.* 2006.

[2] Rittel and Webber 1973, 160–169.

[3] For a discussion of different framings of the climate problem, see Hulme 2009.

solution has been chosen, it is difficult to reverse the effects and the (unintended) consequences.

In addition to these general features of wicked problems, climate change is also characterized by more specific traits that make it a 'super' wicked problem.[4] First, the causes and impacts of, and responses to climate change cut across all sectors of society. Various human activities and sectors of society contribute to increased concentrations of greenhouse gases in the atmosphere, but at the same time may also be affected by the impacts of climate change. Second, if we wish to avoid large-scale, irreversible impacts, an urgent response is necessary. Third, resolving climate change is made difficult by the fact that international and national decision makers cannot fully control the choices of actors who have an impact on addressing climate change. Fourth, climate change is a transboundary problem, and may indeed be 'the greatest collective action problem the international community has yet faced'.[5] This exacerbates tensions between different countries, especially because those who are in the best position to take action have little incentive to do so.[6] Fifth, climate change has an undeniably inter-temporal dimension: to mitigate impacts in the future, action now is needed.[7] Finally, the problem is characterized by various types of scientific uncertainty, including uncertainty regarding the future development of greenhouse gas emissions as well as the impacts (and associated costs) of climate change in the long term.[8]

By its very nature, the climate change problem thus covers a broad range of narrowly defined issue areas, and its resolution inevitably requires a variety of responses. At the international level, this means that the problem is governed by a multitude of regimes with overlapping jurisdictions. It is this fragmented state of global climate governance that forms the point of departure for this book.

2. THE OBJECTIVES OF THIS BOOK

Due to the wickedness of the climate change problem, and given the fragmentation of global climate governance, there are interrelationships

---

4 Levin *et al.* 2007, 4–9; Lazarus 2009, 1159–1183.
5 Cole 2008, 232.
6 Levin *et al.* 2007, 9; Lazarus 2009, 1160.
7 Lazarus 2009, 1174–1176.
8 IPCC 2007, 72–73.

between the climate regime established by the United Nations Framework Convention on Climate Change (UNFCCC) and the wide variety of governance arrangements initiated outside of it. This means that to understand and appraise the totality of global climate governance, it is no longer apposite to focus on the UN climate regime in isolation; instead, the institutional environment of a regime needs to be taken into account.[9] It is necessary to move away from the myopic view of environmental regimes that has been deemed one of the 'intellectual blind spots' in international environmental law scholarship.[10] Tackling this blind spot first of all requires an acknowledgement of the relevance of other institutions in addressing a certain policy problem. Beyond this initial mapping exercise, scholars will also need to examine the nature and consequences of the relationships between the different institutions.

Some degree of overlap between the UN climate regime and non-UNFCCC institutions is likely to be inevitable given the scope of the phenomenon, and may even be necessary for integrated efforts to limit greenhouse gas emissions and adapt to the detrimental impacts of a changing climate. Indeed, the fragmentation of global environmental governance more generally is said to 'reflect the high political salience of environmental issues and their particular problem structure'.[11] Moreover, such overlap can also result in synergies between different elements of a governance architecture. Yet, on a systemic level, the multiplicity of institutions and regimes, and subsequent interactions between them, could also pose a threat to the coherence of global climate governance.

This brings us to the first objective of this book, which is to provide new insights into the *consequences* of the fragmentation of global climate governance and subsequent interactions between different regimes related to climate change. There is a growing number of scholarly contributions on this subject in the field of international relations. These contributions have moved away from an initial focus on typologies of interactions[12] towards more detailed analyses of the consequences of interactions in specific issue areas, and comprise studies on a wide variety of international environmental institutions.[13] International law scholarship has also elaborated on the possible consequences of conflicts between different branches of international law. Although the focus of these

---

[9] Biermann 2007, 332.

[10] E. Fisher *et al.* 2009, 28.

[11] Oberthür and Gehring 2004, 369.

[12] E.g., O.R. Young 1996; Rosendal 2001; Stokke 2001; Oberthür and Gehring 2006a.

[13] See, in particular, the contributions in Oberthür and Gehring 2006b.

studies largely overlaps with the international relations research on institutional interactions, legal studies have been framed primarily by the discussion on the consequences of the 'fragmentation of international law'.[14]

Both bodies of literature have contributed to knowledge on the consequences of interactions between different regimes, but to date there have not been any in-depth studies into the various consequences of regime interactions in the area of global climate governance. There are various detailed case studies examining specific interactions involving the UN climate regime and usually one other issue area (e.g., biodiversity, ozone layer depletion, human rights and trade),[15] but detailed studies involving the climate regime and several other regimes are rare. The literature on the fragmentation of international law is also limited in that its focus has been primarily on international economic law and human rights law,[16] even though some contributions have also taken up the issue in the context of international environmental law[17] and climate change law.[18] Still, there is as of yet no comprehensive study on the consequences of the fragmentation of international climate change law. Furthermore, both international lawyers and international relations scholars have largely overlooked interactions involving non-legally binding agreements, even though their use is becoming increasingly important in global climate governance (as well as global environmental governance more broadly).

Awareness of regime interactions and their consequences raises the possibility of their *management*. The second objective of this book therefore is to examine strategies for dealing with regime interactions in global climate governance. More specifically, I aim to analyse the advantages and drawbacks of different ways of managing interactions, in terms of effectiveness and feasibility.

Such management would generally be aimed at capturing the synergies between different regimes and minimizing potential or actual conflicts. There is a growing body of research on the management of regime

---

[14] ILC 2006.

[15] E.g., Werksman *et al.* 2003; K. Scott 2005; Burns 2009; S. Humphreys 2009; Rosenthal and Watson 2011.

[16] E.g., Cassimatis 2007; Orakhelashvili 2008; van Aaken 2008; Delimatsis 2011.

[17] M.A. Young 2011b; K. Scott 2011; Anton 2012.

[18] Carlarne 2008; van Asselt *et al.* 2008; M.A. Young 2011a.

interactions, especially in global environmental governance.[19] These studies have shed light on the various ways and means of interaction management, as well as the diverse set of actors involved. Existing research also provides a first indication of how legal and political strategies may be suitable for managing different regime interactions,[20] pointing to a potentially complementary role for legal techniques and political approaches in managing regime interactions. However, no study has focused on the variety of formal and informal responses to institutional diversity in one specific issue area.

These questions regarding the consequences and the management of regime interactions are not merely of academic significance. Most states participate in a multitude of regimes and will generally seek to ensure that complying with the commitments under one regime does not undermine compliance with others. Phrased more positively, it makes sense for actors participating in several regimes simultaneously to ensure that activities in one regime contribute to achieving the goals of another. These considerations may be influenced by the pursuit of an overarching goal, such as normative coherence, mutual supportiveness, or sustainable development. They may equally be influenced by the more pragmatic desire to increase efficiency: how to make better use of existing human, financial, and technological resources to implement a variety of agreements. There are at least three challenges for actors participating in different regimes in this regard: (1) to comply with commitments stemming from various regimes; (2) to actively participate in different regimes; and (3) to monitor and report the implementation of various agreements.[21] Hence, there are genuine reasons for regime participants, especially from developing countries, to be concerned about the fragmentation of global climate governance, and to find ways of coping with the variety of institutional arrangements at the international level.

3. AN INTERDISCIPLINARY APPROACH

This book adopts an interdisciplinary, problem-driven approach, taking into account both legal and political aspects of global climate governance. The approach therefore includes elements of international law and

---

19 Oberthür 2009; Oberthür and Stokke 2011b; van Asselt 2011b; Zelli 2011b.

20 Wolfrum and Matz 2003.

21 Anton 2012.

international relations research.[22] The attraction and rationale of this interdisciplinary approach lies in its ability to address questions that go beyond the international lawyer's comfort zone. For international lawyers, the proliferation of sites of governance, the increasing presence of non-state actors in global governance or the use of formal or informal means of international cooperation are not so much phenomena that need to be explained, but rather lead to descriptive and prescriptive questions about the role of international law vis-à-vis such developments. In the context of this book, this has led to questions about the up- and downsides of the fragmentation of international law (e.g., how can international law avoid or resolve conflicts between different norms and treaties?); the deformalization of international law (e.g., what is the status of 'soft' law under international law?); and the rise of global administrative law (e.g., how can international law deal with autonomous activities of international organizations and other non-state actors, and secure their accountability and legitimacy?). Although international legal scholars thus certainly have sought to deal with the new phenomena that have puzzled international relations scholars, these debates are taking place on a rather systemic level, reflecting the uncertain nature of what international law is (and should be) today.

An interdisciplinary approach presents some inherent challenges. In the case of work bridging the gap between international relations and international law, it should be kept in mind that there is no single school of thought that represents the whole of either discipline.[23] Furthermore, it is imperative to pay attention to the specificities of both disciplines. For international lawyers, this means it is important to identify what makes international law 'law'. From a practical perspective, an added challenge is to reconcile often-diverging vocabularies.[24] For instance, the concepts of 'regime' and 'institution' do not always have the same meaning in both disciplines.[25] Similarly, experts in different disciplines may examine the same phenomena but use diverging vocabulary to describe them.[26]

---

22 On the relationship between international law and international relations, see Abbott 1989; Slaughter-Burley 1993; Keohane 1997; Slaughter *et al.* 1998; Beck 2009; Hafner-Burton *et al.* 2012; Dunoff and Pollack 2013a; Dunoff and Pollack 2013b.

23 Klabbers 2005, 37–38.

24 Cf. E. Fisher *et al.* 2009, 21.

25 M.A. Young 2011b, 19–20.

26 For a critical analysis of this development, see Koskenniemi 2009, 406–410.

I would not go as far as to say that the pursuit of interdisciplinarity in the case of international law and international relations boils down to an 'American crusade' that 'cannot but buttress the justification of American hegemony in the world', as Martti Koskenniemi put it provocatively.[27] It is not (or is at least no longer) possible to capture international relations research in simple 'America *versus* the rest of the world' dichotomies.[28] Moreover, even though American-dominated (neo-)realism may still be a major school of thought in international relations, other schools of thought (e.g., institutionalism and constructivism) that provide sufficient counterbalance have emerged over time.[29] Furthermore, the dialogue with international relations scholars forces international lawyers to more clearly argue the distinguishing characteristics of (international) law, and could therefore also strengthen the discipline of international law.[30] I argue that interdisciplinary research that keeps its limitations in mind offers a fruitful way of exposing different angles of the same problem.

In the context of this book, the discipline of international law provides valuable insights on determining the existence of a conflict between norms; examining the possibilities for addressing such conflicts through the law of treaties; and identifying the scope for legal techniques to address conflicts and enhance synergies between international agreements. The discipline of international relations helps by, among other things, shedding light on the driving forces behind regime interactions; identifying causal mechanisms for interactions; and assessing the impacts of interaction management on regime effectiveness. Both disciplines address different, yet related, questions and can thus present complementary insights and lessons for each other.

4. CASE STUDIES

A methodology needs to be 'best suited' for the type of question asked.[31] The methodology for this book is to a large extent empirically driven and follows a case study approach. The primary reason for using case studies is to provide insights into the different types of interactions and the different types of interaction management. The three case studies thereby

---

27 Koskenniemi 2000, 29–30.
28 Beck 2009, 21–22.
29 Beck 2009, 18.
30 Hurrell 2000, 332–333.
31 E. Fisher *et al.* 2009, 15.

allow for a comprehensive analysis of the various forms of regime interaction in global climate governance.

For the first objective (the consequences of regime interactions) I analyse various interactions between climate-related regimes with a view to determining the effects of interactions on the 'target' regimes. The interactions are analysed according to the interacting objects, the causal mechanisms driving interactions, intentionality and the effects of interactions.

For the second objective, I examine the effectiveness and feasibility of legal techniques and institutional coordination to address regime interactions. To this end, I identify distinguishing characteristics of the three case studies to determine how legal techniques and institutional coordination can complement each other, and how conflicts can be avoided and resolved and synergies enhanced. In particular, employing the emerging body of literature on 'interplay management',[32] I draw comparisons between the management efforts in each of the cases to identify opportunities and limitations of different ways of interaction management.

The three cases are selected from the wider universe of international regimes in the area of climate change. The size of the issue area of 'climate change' depends on how one draws the boundaries around it, an inherently subjective exercise. Notwithstanding this subjectivity, it is evident that several international institutions outside of the UNFCCC are of clear and immediate relevance to addressing climate change (either from the perspective of reducing emissions or adapting to the impacts of climate change). These institutions include not only regimes that directly aim to tackle climate change and regimes that concern environmental problems physically linked to climate change, but also other regimes that may have non-environmental policy objectives but which can still be considered as highly important for responding to the climate problem, including regimes in the areas of international trade and investment, human rights, agriculture and security. While one can also construct links between climate change and, to name but a few issue areas, children's rights, maritime piracy, space-faring, telecommunications and terrorism, these links will tend be less direct and the interactions are likely to be less visible. So while I do not generally discard the relevance of any particular regime for climate change (and vice versa), my focus is on three cases that belong to a subset of international regimes that are of

---

[32] Oberthür 2009; Stokke and Oberthür 2011.

clear and immediate relevance for addressing climate change. Moreover, given the different problem structures of climate change mitigation and adaptation, the book focuses primarily on regime interactions related to climate change mitigation.

The three cases each represent a different sphere of global climate governance.[33] They thereby not only display the breadth of regime interactions, but also illustrate the different ways in which interactions have been or can be managed. Moreover, in each case there have been contentious interactions, which have drawn significant political and academic attention and spurred rule development at different levels of governance. This, in turn, means that there is a significant empirical base to draw from.

The first case addresses the *institutional diversity within the issue area of international climate change law and governance*, looking at interactions between regimes that explicitly aim to combat climate change: the multilateral climate regime, established by the UNFCCC and the Kyoto Protocol, on the one hand, and the Asia–Pacific Partnership on Clean Development and Climate (APP) as an example of a 'minilateral' regime promoting clean technologies on the other. The APP is representative of a number of non-legally binding governance arrangements that have been initiated outside the UN framework, consisting of only a limited number of participating countries. Initial studies indicated that the APP may have detrimental effects. The research complements these studies by relating the interactions to the legal nature of the UNFCCC (as 'hard law') and the APP (as 'soft law').

The second case relates to the *institutional diversity within international environmental law and governance*, examining the interactions between the UN climate regime and the biodiversity regime established by the Convention on Biological Diversity (CBD). The CBD is part of the increasingly autonomous subsystem of international environmental law,[34] and shares important principles (e.g., the principle not to cause transboundary harm) and instruments (e.g., national reporting systems) with the climate regime. Given these commonalities, one would anticipate little conflict between the climate and biodiversity treaties and rather expect to find potential for synergies. Nevertheless, existing literature has identified potential conflicts between the implementation of the climate treaties and the objectives of the CBD. The book builds on these studies by examining regime interactions related to forest carbon.

---

[33] Cf. Zelli 2011a.

[34] E.g., Bodansky 2006; Bodansky *et al.* 2007.

The third case tackles *overlaps between different branches of international law and governance*, examining interactions between the World Trade Organization (WTO) and the climate regime. The international trade regime has been chosen for its suitability as an example of how regimes with very different objectives and principles interact. The relationship between climate change and international trade law has received increasing attention, focusing mainly on potential conflicts. I expand on this literature by analysing the variety of regime interactions at the climate–trade nexus, with a focus on unilateral climate-related trade measures.

5. SOME LIMITATIONS

Like any product of research, this book is not without its limitations. First of all, I am concerned with law and governance at the *international* level rather than the national or sub-national level. Consequently, this book deals with questions of 'horizontal' rather than 'vertical' interactions.[35] This is not to deny the importance of other levels of governance. As scholars working on multi-level governance will be quick to point out, the different levels of governance are intrinsically related.[36] I also do not wish to downplay the importance of implementation processes at the domestic level in managing overlapping commitments from different agreements.[37] However, in line with regime theorists,[38] I am primarily interested in the role of international regimes and particularly their interrelationships. Furthermore, the role of governments as well as non-state actors is highlighted in discussions of 'autonomous interaction management', in which I examine interaction management beyond the interacting regimes.

Second, while I acknowledge the importance of enhancing understanding of the wide variety of transnational governance arrangements that have emerged beyond the state (and beyond the individual regime), this book is first and foremost concerned with interactions of regimes mainly created by states, for states. Although new forms of private and public–private governance have emerged in recent decades, I argue that regimes created by states continue to play a central role in international law and governance. Nevertheless, the book touches upon the role of non-state

---

35 O.R. Young 2002, 83–138.

36 See, e.g., Hooghe and Marks 2003; Cash *et al.* 2006.

37 Velasquez and Piest 2003.

38 E.g., Levy *et al.* 1995; Hasenclever *et al.* 1997; O.R. Young 1999a.

actors in the case study chapters, notably by discussing their role in managing regime interactions.

A third caveat, which holds for each of the case studies to at least some extent, is that this book does not present a purely *ex post* view of regime interactions, in contrast to most other studies.[39] The simple observation that a regime interaction has not (yet) taken place does not mean that its analysis is irrelevant or impossible. Interactions themselves are often 'moving targets' that may only manifest themselves over time.[40] This calls for attention to *potential* interactions. The risk of discussing potential interactions is that it could lead to (over)speculation. One could argue, for instance, that the identified potential effects of the APP on the UNFCCC are irrelevant, as the APP has ceased to exist (see Chapter 6). Such a criticism, however, would disregard possible lessons that can be learned. More specifically, the case study could shed light on the conditions under which non-UNFCCC institutions might be in conflict with, or enhance synergies with, the climate regime.

Finally, I do not pretend to offer ideal-type solutions for the various regime interactions. As I argue in this book, doing so would require the definition of an objective standard against which solutions could be evaluated. However, for super wicked problems like climate change there is unlikely to be such an objective standard. And even if it could somehow be objectively derived, a solution would likely lead to unexpected and possibly unwanted ripple effects. Nevertheless, I seek to provide practical recommendations on how state and non-state actors could address specific interactions identified in the case study. Furthermore, the conceptual insights from this book go beyond individual cases of interactions, and provide more in-depth knowledge about the usefulness of specific forms of interaction management.

6. PLAN OF THE BOOK

This part of the book proceeds with an overview of the global response to the climate change problem, in which I show how the UN climate regime has increasingly been flanked by other governance arrangements (Chapter 2).

Building on the relevant literature in international law and international relations, Part II of the book presents the analytical framework employed in the subsequent chapters. To place the conceptual discussions in the

39 E.g., the contributions in Oberthür and Gehring 2006b.

40 Zelli 2011b, 202.

book in a wider context, I first introduce the concept of 'fragmentation' to describe the situation of multiple overlapping regimes and critically discuss various interpretations of the concept. Furthermore, I examine whether the phenomenon of fragmentation should be seen as benign or not (Chapter 3). I then move on to show how the debate on fragmentation in international law is essentially connected to discussions on institutional interactions among international relations scholars. I distinguish various types of regime interactions and zoom in on the consequences of such interactions, introducing the notions of 'conflict' and 'synergy' used in the book (Chapter 4). In a further step, I introduce strategies for dealing with regime interactions under the heading of 'interaction management'. I discuss various types of legal techniques for avoiding or resolving conflicts and provide an overview of political approaches to interaction management, in particular institutional coordination (Chapter 5).

Part III comprises the three case studies in which this framework is applied. First, I examine the interactions between the UN climate regime and minilateral clean technology agreements, focusing on interactions with the 'soft law' APP (Chapter 6). The next chapter focuses on the interactions between the climate treaties and the CBD, discussing in particular the pertinent and policy-relevant interactions related to forests (Chapter 7). In the third case study I investigate the interactions between the climate regime and the world trading system, focusing on the highly controversial issue of climate-related trade measures at the border (Chapter 8).

To put the case studies back in perspective, Part IV starts with a cross-cutting synthesis and analysis of the consequences and management of regime interactions in global climate governance (Chapter 9). In the final chapter, I present the main conclusions and discuss ways forward from the perspectives of policymaking and further academic inquiry (Chapter 10).

2. The evolution of global climate governance

1. INTRODUCTION: THE SCOPE OF THE CHALLENGE

There is a solid scientific basis for international action to mitigate the causes and impacts of climate change. As the Fifth Assessment Report of the Intergovernmental Panel on Climate Change (IPCC) indicated, we know with more than 95 per cent certainty that human activities contribute to an increase in the global average temperature.[1] Emissions of greenhouse gases, particularly carbon dioxide (CO_2), provide a major contribution to the warming trend and it is evident that greenhouse gas concentrations are increasing. CO_2 concentrations have increased from 280 parts per million (ppm) in pre-industrial times[2] to over 400 ppm in 2013.[3] CO_2 is primarily emitted through the use of fossil fuels, although land-use changes (such as deforestation) also play an important role.

Climate science also provides mounting evidence of the impacts of climatic changes. The increasing temperatures are expected to lead to impacts across the globe, some of which will be and are being felt worldwide (e.g., sea level rise[4]) whereas others (e.g., extreme weather events) will vary for different regions. How climate impacts will be felt by humans and ecosystems depends to a large extent on their climate vulnerability and their ability to adapt to climate change. It is clear that many climate impacts are very unevenly distributed, and that the least developed countries are at the same time the most vulnerable and have the lowest capacity to adapt.

Although a state of the art on the responses to climate change is invariably influenced by value judgements of what the appropriate response should be (or, as I suggested in Chapter 1, on how one frames the problem), a few general remarks can be made here. First, both

---

1 IPCC 2013, 15.

2 IPCC 2007, 37.

3 In 2013, the symbolic value of 400 ppm was reached (and exceeded). See BBC 2013.

4 IPCC 2007, 45.

national governments[5] and various scientists[6] have embraced the objective to keep the increase of the global average temperature below 2°C relative to pre-industrial times, with countries vulnerable to climate impacts calling for limiting temperature increases to 1.5°C.[7] The wide adoption of this objective provides a good indication of the 'destination' of international climate policy.[8] Second, translating this destination into greenhouse gas stabilization targets and, eventually, greenhouse gas emission reduction targets, although fraught with uncertainties, increasingly provides the international community with an idea of the task ahead of us. Scientists have provided a clear indication that keeping temperature increases below 2°C with a considerable degree of certainty (over 93 per cent) would require that greenhouse gases are stabilized at 350 ppm CO_2-equivalent.[9] Furthermore, assessments show that to stay below 2°C with over 75 per cent certainty, it would be necessary to limit cumulative CO_2 emissions between 2000 and 2050 to 1000 gigatonnes.[10] Approximately half of this trillion tonne budget has already been emitted,[11] meaning that at the rate CO_2 is currently being emitted, net emissions would need to be zero around 2050.[12] Keeping temperature increases below 1.5°C will be even more challenging: none of the existing climate change mitigation scenarios suggest how we can achieve that goal in this century.[13]

To achieve the objectives adopted by the international community and to avoid climate impacts that would be perceived as dangerous by many countries and individuals, a drastic cut in emissions is therefore not only necessary, but also needs to happen very urgently.[14] What follows is a short introduction of how the international community has sought to achieve this.

---

[5] UNFCCC Decision 1/CP.16, para. 4.

[6] E.g., Azar and Rodhe 1997; O'Neill and Oppenheimer 2002, 1972; Mann 2009.

[7] E.g., Decision 1/CP.16, para. 4.

[8] Gupta and van Asselt 2006, 83.

[9] Meinshausen 2006, 270.

[10] Meinshausen *et al.* 2009, 1158.

[11] Allen *et al.* 2009a, 1163.

[12] Allen *et al.* 2009b, 57.

[13] Rogelj *et al.* 2011, 414.

[14] Anderson and Bows 2011, 41; Rogelj *et al.* 2011, 413.

2. CRAFTING A GLOBAL RESPONSE: A SHORT HISTORY OF THE UN CLIMATE REGIME

2.1 The United Nations Framework Convention on Climate Change

The climate problem gained prominence on the international political agenda in the 1980s, in part due to the emerging scientific consensus, and in part because of increased public awareness.[15] These and other factors led to the adoption of the United Nations Framework Convention on Climate Change (UNFCCC) in 1992 at the United Nations Conference on Environment and Development in Rio de Janeiro. The Convention aims to achieve 'stabilization of greenhouse gas concentrations in the atmosphere at a level that would prevent dangerous anthropogenic interference with the climate system',[16] a goal that, as noted above, has been translated into avoiding temperature increases of more than 2°C above pre-industrial levels.

The Convention puts forward several principles,[17] including those of inter-generational and intra-generational equity, common but differentiated responsibilities and respective capabilities of developed and developing countries,[18] the need for a precautionary approach,[19] the right to sustainable development[20] and the promotion of a supportive, open economic system.[21]

The principle of common but differentiated responsibilities and respective capabilities implies that all countries bear a responsibility for dealing with global problems such as climate change. The UNFCCC goes on to add that these responsibilities need to be differentiated on the basis of countries' varying historical responsibility for the problem, as well as their capacity to deal with climate change. The principle is related to the leadership that is expected of developed countries in climate change abatement.[22] The leadership paradigm implies that Annex I (developed) countries should lead in reducing greenhouse gas emissions and provide non-Annex I (developing) countries with financial assistance and the necessary clean technologies to meet the agreed incremental costs of

---

15 Bodansky 1993, 458–461.
16 UNFCCC, Art. 2.
17 Bodansky 1993, 501–502.
18 UNFCCC, Art. 3.1.
19 UNFCCC, Art. 3.3.
20 UNFCCC, Art. 3.4.
21 UNFCCC, Art. 3.5.
22 UNFCCC, Art. 3.1.

implementation.[23] Developing countries were opposed to any quantified emission limitation or reduction obligations from the outset, but they nevertheless took on general commitments that apply to all parties. Article 4 first sets out the commitments for all parties, such as the development of national inventories of anthropogenic greenhouse gas emissions by sources and removals by sinks,[24] the formulation and implementation of programmes containing climate change mitigation measures,[25] consideration of climate change in relevant social, economic and environmental policies and actions,[26] and scientific and technological cooperation.[27] It goes on to specify further commitments for the Annex I countries, requiring them to adopt policies and measures to limit their greenhouse gas emissions and enhance their sinks and reservoirs with the aim of returning individually or jointly to their 1990 emission levels by 2000.[28] Article 4 also gives an elaboration of the leadership paradigm, by stating that developed countries 'shall provide such financial resources, including for the transfer of technology … to meet the agreed full incremental costs of implementing measures'.[29] The importance of these Annex I party obligations is underlined by the provision that implementation of the commitments by non-Annex I countries is made conditional on 'the effective implementation by developed country parties of their commitments under the Convention related to financial resources and transfer of technology'.[30]

The UNFCCC has been widely ratified, including by all major emitters, and launched an ongoing international negotiation process. This in itself should not be underestimated. As Depledge and Yamin note, '[t]he negotiating environment of a regime enmeshes delegations in a dense web of meetings, practices, processes, and rules, generating an inherent motivation among negotiators to advance the issue'.[31] This 'momentum' ensures that all countries have moved forward (albeit sometimes painfully slowly and incrementally) towards a common goal of avoiding dangerous climate change.

---

23 Gupta 1998, 183–186.
24 UNFCCC, Art. 4.1(a).
25 UNFCCC, Art. 4.1(b).
26 UNFCCC, Art. 4.1(f).
27 UNFCCC, Art. 4.1(g).
28 UNFCCC, Art. 4.2(a) and (b).
29 UNFCCC, Art. 4.3.
30 UNFCCC, Art. 4.7.
31 Depledge and Yamin 2009, 439.

2.2 The Kyoto Protocol

Even though the UNFCCC was a crucial first step in global climate change abatement efforts, it lacked teeth in the form of legally binding quantified emission limitation and reduction obligations for developed countries. This deficiency was acknowledged at the first Conference of the Parties (COP) held in Berlin in 1995, where parties agreed to 'take appropriate action beyond 2000, including the strengthening of the commitments of Annex I Parties … through the adoption of a protocol or another legal instrument'.[32]

The adoption of the Kyoto Protocol[33] in 1997 (and its subsequent implementation) reinforced the impression that the UN process was the 'only game in town'[34] or at least the main show. Most importantly, the Protocol introduced legally binding greenhouse gas emission targets for all industrialized countries. Countries listed in Annex B (corresponding to a large extent with Annex I of the UNFCCC) committed to reduce net greenhouse gas emissions by 5.2 per cent compared to 1990 levels between 2008 and 2012.[35] Like the UNFCCC, the Protocol does not require developing (non-Annex B) countries to reduce or limit their emissions. The Protocol provides a non-exhaustive list of policies and measures that Annex B parties can implement to achieve the Kyoto targets, including energy efficiency measures, the protection and enhancement of sinks and reservoirs, sustainable forestry practices, sustainable agriculture, research on and promotion of the use of renewable energy, reducing and phasing out market imperfections, and measures in the transport sector.[36] Furthermore, the Protocol introduced three flexibility mechanisms to assist developed countries in achieving their targets: Joint Implementation,[37] the Clean Development Mechanism (CDM)[38] and international emissions trading.[39] The carbon market established by the CDM is a key achievement of the Kyoto Protocol. Between 2002 and 2008, 1.9 billion credits worth US$ 23 billion were contracted, which in turn could leverage up to US$ 106 billion in carbon finance.[40]

32 UNFCCC Decision 1/CP.1, preamble.
33 Grubb *et al.* 1999; Oberthür and Ott 1999; Depledge 2000.
34 Thompson 2001.
35 Kyoto Protocol, Art. 3.
36 Kyoto Protocol, Art. 2.1(a).
37 Kyoto Protocol, Art. 6.
38 Kyoto Protocol, Art. 12.
39 Kyoto Protocol, Art. 17.
40 Kossoy and Ambrosi 2010, 42.

While the CDM has received its fair share of criticism,[41] it is undeniable that it has helped to scale up low-carbon investments in developing countries.

The Kyoto Protocol was a major step forward in terms of specifying targets and timetables for Annex B countries, but it left many important details for future negotiations. In particular, the COP was mandated to further elaborate on the design of the Protocol's flexibility mechanisms and its compliance procedure. Moreover, after the adoption of the Kyoto Protocol, the United States (US) argued that the lack of quantitative commitments for key developing countries, such as China and India, did not meet the US Senate's demand for meaningful action by developing countries.[42] Even though the US had signed the Protocol, it was far from clear whether it would ratify the agreement. In the face of this uncertainty, other developed countries were also reluctant to ratify the treaty.

Disagreement over several contentious issues related to operationalizing the Kyoto Protocol's provisions culminated in the failure of COP6 in 2000.[43] Following the failed COP, the new US Administration rejected the Kyoto Protocol in 2001 as an agreement that was 'fatally flawed in fundamental ways'.[44] After the United States' withdrawal, the European Union (EU) sought to encourage ratification by Russia and Japan to ensure that the treaty would enter into force. At the resumed COP6bis in 2001, the result was a political agreement on many of the outstanding issues that resulted in the failure a year before. This agreement was fleshed out in the Marrakech Accords, adopted later that year. The Accords paved the way for the entry into force of the Kyoto Protocol, which was secured after a bargain was struck between Russia and the EU, which included trade concessions under the WTO.[45]

2.3 Towards a Post-2020 Climate Agreement

Following the Marrakech Accords, climate negotiators' attention gradually shifted towards implementation of the Protocol, and agreements were reached on further details regarding the Protocol's flexibility mechanisms and its financial mechanisms. With the Kyoto rulebook largely in place, and implementation having begun in most countries, the question of how

---

41 E.g., van Asselt and Gupta 2009.
42 S. Res. 98, 105th Cong. (1997) (enacted).
43 Grubb and Yamin 2001.
44 White House 2001.
45 Douma 2006, 61–62.

to proceed beyond Kyoto's first commitment period (2008–2012) became a key issue.

In 2005, parties made a first set of small steps to discuss the future of international climate policy.[46] In the context of the UNFCCC, parties reached agreement on starting an open, non-binding dialogue.[47] Discussions on new commitments for developed countries were also initiated on the basis of the Kyoto Protocol,[48] leading to the establishment of the Ad Hoc Working Group on further commitments for Annex I Parties under the Kyoto Protocol (AWG-KP).[49] Two years later, parties adopted the 'Bali Action Plan', which launched 'a comprehensive process to enable the full, effective and sustained implementation of the Convention through long-term cooperative action, now, up to, and beyond 2012, in order to reach an agreed outcome and adopt a decision at its fifteenth session'.[50] These negotiations were to take place within the Ad Hoc Working Group on Long-term, Cooperative Action (AWG-LCA).

The AWG-LCA's mandate instructed it to reach agreement by the time of COP15 in Copenhagen in 2009. Before the Copenhagen summit, it became increasingly evident that a legally binding agreement would be an unlikely outcome, and that a set of detailed political decisions would be the best possible result. Not only had the negotiations become incredibly complex, it was also evident that some of the crunch issues (including mitigation commitments and actions by developed and developing countries, and financial support for developing countries) would be very difficult to resolve. While this development lowered expectations, the stakes were raised by the presence of more than 100 heads of government at the summit.

The Copenhagen COP certainly did not live up to the high expectations.[51] The Copenhagen Accord adopted in the Danish capital was not a legally binding international agreement, but rather a political declaration by a limited number of countries. While the Accord encouraged developed countries to list quantified economy-wide targets and requested developing countries to list their nationally appropriate mitigation actions,[52] it also contained various open-ended and ambiguous statements of good intentions.

---

[46] Bausch and Mehling 2006.
[47] UNFCCC Decision 1/CP.11, para. 2.
[48] Kyoto Protocol, Art. 3.9.
[49] Kyoto Protocol Decision 1/CMP.1, para. 2.
[50] UNFCCC Decision 1/CP.13, para. 1.
[51] Bodansky 2010; Rajamani 2010.
[52] Draft UNFCCC Decision -/CP.15, paras. 4–5.

One year later, negotiators in Cancún had the difficult task of keeping the UNFCCC process alive. To the surprise of many observers, they managed to do so by adopting the Cancún Agreements,[53] which fleshed out the Copenhagen Accord. While the Agreements lack the ambition and the specificity to achieve meaningful greenhouse gas emission reductions,[54] they brought country emission reduction pledges made in the context of the Copenhagen Accord within the formal UNFCCC framework through a COP decision. Furthermore, they added details on the agreement on monitoring, reporting and verification for mitigation commitments and actions for developed and developing countries; established a new funding mechanism (the Green Climate Fund) and a new Technology Mechanism; and incorporated an agreement on reducing emissions from deforestation and forest degradation. The Cancún summit thus seemingly (re)affirmed the central importance of the UNFCCC process. At the very least, the summit is evidence of its resilience.

The direction in which the UNFCCC process will head next is in principle still open, although there are some clear signs. Parties at COP17 in Durban reached agreement on further negotiations, through the newly created Ad Hoc Working Group on the Durban Platform for Enhanced Action (ADP),[55] but the legal form of a new climate agreement remains uncertain, as the negotiations are to lead to 'a protocol, another legal instrument or an *agreed outcome with legal force* under the Convention applicable to all Parties' by 2015.[56] It is notable that the new agreement will only 'come into effect and be implemented from 2020'.[57] In the meantime, the Kyoto Protocol will continue to be in force, with parties reaching agreement on an amendment for a second commitment period in Doha in 2012.[58] Acknowledging the need for non-Kyoto parties to take mitigation action before 2020, the ADP negotiations incorporate a separate 'work stream' discussing pre-2020 action.[59]

The path chosen through the Copenhagen Accord and Cancún Agreements is one that moves away from the 'targets-and-timetables' approach introduced by the Kyoto Protocol, towards a system of self-selected

---

53 UNFCCC Decision 1/CP.16.

54 Rogelj *et al.* 2010, 1126.

55 UNFCCC Decision 1/CP.17, para. 2; see Rajamani 2012.

56 UNFCCC Decision 1/CP.17, para. 4 (emphasis added).

57 UNFCCC Decision 1/CP.17, para. 4.

58 UNFCCC Decision 1/CMP.8. Annex B countries like Canada (which had already withdrawn from the Kyoto Protocol), Japan and Russia have made it clear they do not intend to participate in the second commitment period.

59 UNFCCC Decision 1/CP.17, para.

mitigation commitments and actions accompanied by international monitoring, reporting and verification procedures. Under this new approach, differentiation between developed and developing countries, a cornerstone of both the Convention and the Protocol, moved to the background and is being replaced by a system of differentiation between all countries – irrespective of whether they are developed or developing.[60] A key question for the 2015 agreement, and for the architecture of global climate governance more broadly, will be the extent to which the 'bottom-up' system of national mitigation pledges will be accompanied by 'top-down' elements that track progress towards a global objective.

3. THE FRAGMENTATION OF GLOBAL CLIMATE GOVERNANCE

As the overview so far shows, the initial development of global climate governance in the 1990s can be characterized by the central role played by the UNFCCC. Despite criticisms voiced about the UNFCCC process[61] and a turbulent period following the United States' withdrawal from Kyoto, the process is still alive and kicking. However, claims that the UNFCCC is the main show in town should be critically examined in light of climate action emerging at various levels and jurisdictions and undertaken by a variety of public and private actors.

In the 2000s, attention was increasingly drawn to initiatives outside of the UNFCCC context. In part, this can be explained by the US exit from Kyoto. However, the emergence of new governance arrangements (and the shifting focus of existing ones) is also due to the higher profile of climate change on the international policy agenda. Without pretending to provide an exhaustive list of non-UNFCCC governance arrangements, it is possible to distinguish several broad categories.

First of all, existing international organizations, such as the World Bank[62] have sought to integrate climate change concerns in their operations. In addition, various international environmental regimes have started to address climate change-related issues falling within their mandates. For instance, from the late 1990s onwards, parties to the CBD have adopted numerous decisions highlighting the links between biodiversity and climate change (see Chapter 7). As another example, the Montreal Protocol, which is aimed at the reduction of ozone-depleting

---

60 Rajamani 2011, 519.

61 E.g., Victor 2011; Prins and Rayner 2007.

62 World Bank 2008a.

substances, has even been argued to be more successful in terms of reducing global greenhouse gas emissions than the Kyoto Protocol.[63] Furthermore, certain climate policies may fall within the remit of existing international regimes even if such regimes are not explicit about the potential interactions. For instance, certain forms of oceanic carbon sequestration, a potential form of climate change mitigation, may be in violation of the law of the sea;[64] bilateral or regional investment agreements may prohibit the kind of conditioning of investments that the Kyoto Protocol's flexibility mechanisms promote;[65] certain climate policies may result in discrimination between countries in a way that contravenes international trade law (see Chapter 8); and so on.

Other initiatives comprise high-level, club-like forums involving the political leaders of a limited number of important countries, such as various Group of 8 (G8) summits held since 2005, and the Group of 20 (G20) meetings held since 1999.[66] Another initiative is the Major Economies Process on Energy Security and Climate Change launched by US President Bush in 2007, which has been continued as the Major Economies Forum by President Obama.[67] Yet other approaches have sought to bring together climate change negotiators in less contentious, informal settings. An example is the Cartagena Dialogue organized in the run-up to the Cancún climate summit (and still reconvenes regularly), which was credited with contributing to a positive negotiating atmosphere.[68] These mostly novel 'minilateral'[69] initiatives tend to focus largely on the major greenhouse gas emitters, although increasingly suggestions are being tabled to bring developing countries with a high stake in climate politics, such as small island states, least-developed countries and oil-producing nations, into the fold.[70]

Yet other governance arrangements have taken the shape of multi-stakeholder partnerships involving governments, corporations and/or non-governmental organizations. These partnerships may focus on particular technologies, such as the Carbon Sequestration Leadership Forum, the Global Methane Initiative and the International Partnership for Hydrogen and Fuel Cells in the Economy; or support investment and policy

---

63 Velders *et al.* 2007, 4814.
64 K. Scott 2005.
65 Werksman *et al.* 2003.
66 Bausch and Mehling 2011, 26–33.
67 Bausch and Mehling 2011, 23–25.
68 Bausch and Mehling 2011, 34–35.
69 Naím 2009.
70 Eckersley 2012.

development for several technologies, such as the (now-defunct) APP and the Renewable Energy and Energy Efficiency Partnership (see Chapter 6).

Another category consists of the various regulated and voluntary carbon markets that have been established after the adoption of the Kyoto Protocol.[71] These include large regulatory markets such as the EU emissions trading system. While this trading system is now formally linked to the Kyoto Protocol's flexibility mechanisms, it was created with an expectation that it could function independently from the international legal context at a time when it was still unclear whether the Protocol would enter into force.[72] Even more independent from the UNFCCC process are the voluntary carbon markets, which have emerged in particular to cater to the demand of companies and individuals to offset their emissions.[73] The creation of regulatory and voluntary markets has in turn led to the emergence of new arrangements that seek to govern these markets. For example, voluntary standards for carbon offsetting such as the Voluntary Carbon Standard and the Gold Standard have been created to ensure some level of oversight of the voluntary markets in the absence of (international) regulatory bodies.[74]

Other relevant initiatives undertaken by non-state actors include actions to hold corporations to account for their carbon footprints, either through self-regulation (e.g., the Carbon Disclosure Project) or through scrutiny by civil society organizations.[75] Moreover, private actors that are specifically affected by climate change, such as the insurance industry, have started to respond to the emerging risks by autonomously taking measures.[76]

Finally, numerous sub-national efforts have been launched in recent years. Especially in the US, where policymakers have struggled to put in place meaningful climate policies at the federal level, states and other sub-national actors have become increasingly active.[77] California, for instance, adopted the Global Warming Solution Act in 2006, capping its emissions at 1990 levels by 2020, and setting the additional objective of reducing emissions by 80 per cent by 2050. Furthermore, several states have become involved in (transnational) emissions trading systems, such

---

[71] Bernstein *et al.* 2010.
[72] Biermann *et al.* 2009.
[73] Pattberg and Stripple 2008.
[74] Lovell 2010.
[75] Pattberg and Stripple 2008, 383.
[76] Jagers and Stripple 2003.
[77] Carlarne 2010, 61–97.

as the Regional Greenhouse Gas Initiative. Other sub-national initiatives have involved municipal governments. In addition to stand-alone initiatives at the local level, these also include the creation of transnational networks through which urban actors have sought to cooperate on climate change issues, such as the Cities for Climate Protection campaign.[78]

In summary, global climate governance in the 21st century has increasingly emanated from different sources, leading several scholars to point to an emerging 'regime complex' for climate change.[79] With others, I have argued that the current state of global climate governance can be described as 'fragmented'.[80] Some of the initiatives can be seen as a response to the UNFCCC process (either in support of it, or promoting alternative discourses), whereas others have seemingly emerged in an autonomous fashion. Although governance arrangements outside of the UNFCCC may come and go, it is nevertheless likely that a variety of initiatives outside of the UN climate negotiations will persist.

4. CONCLUSION

This chapter has provided a brief overview of the state of climate science, showing that to achieve the objectives set by the international community, drastic and urgent emission cuts will be necessary. The major response to this challenge by the international community has been the creation of the UNFCCC in 1992. The Convention provides broad guidance for countries by introducing the objective of avoiding dangerous anthropogenic interference with the climate system and by outlining several principles, including notably the principle of common but differentiated responsibilities and respective capabilities. The Kyoto Protocol adds flesh to the Convention by setting targets for industrialized countries and by introducing various mechanisms to achieve these targets, but the negotiations after its adoption show that it has not been easy to align the interests of the different countries and country coalitions participating in the climate regime.

Although the UN climate regime has played a pivotal role in global climate governance, particularly in the 1990s, it is not the only response to the climate challenge. National-level and non-state climate change initiatives have been in place since then, and other international institutions have taken up a much more prominent role, especially since the

---

78 Betsill and Bulkeley 2004.

79 Keohane and Victor 2011; Abbott 2012.

80 Biermann *et al.* 2009.

early 2000s. This increasingly fragmented governance landscape is signified by a complex web of institutions, with various individual elements of global climate governance interacting with each other. It is these interactions that are the subject of the remainder of this book.

PART II

Analytical framework

3. The concept of fragmentation

1. INTRODUCTION

Both international lawyers and international relations scholars have been preoccupied with the quest for coherence in international law and global governance. International legal scholars have primarily sought to establish formal hierarchical relations between different international legal instruments, whereas international relations scholars have mainly focused on explaining how, why and with what consequences institutions interact. Drawing on the insights from both disciplines, this chapter sets the scene for an analytical framework for addressing questions raised by institutional diversity in international law and governance. This analytical framework will be outlined in more detail in the subsequent two chapters.

The chapter focuses on 'fragmentation' as a concept to describe and analyse the institutional landscape of global governance characterized by multiple overlapping regimes. It highlights the contentious aspects of the term and discusses different types of fragmentation that can be discerned. In addition, it provides a theoretical exploration of the consequences of fragmentation, showing how scholars in international law and international relations have suggested that the fragmentation of international law and governance may be beneficial as well as detrimental.

2. THE CHALLENGE OF DEFINING FRAGMENTATION

The use of the term 'fragmentation' is based on a conscious choice, as I explain in this section. Yet the very use of this word has been the subject of a vigorous debate, especially among international lawyers.[1] The term came to the forefront in 2000 when it was included in the work programme of the International Law Commission (ILC).[2] The discussions

[1] The term also has also been used in the international relations literature. See, e.g., Bernstein and Ivanova 2007; Biermann *et al.* 2009.

[2] ILC 2000, para. 729.

in the ILC and its final report submitted in 2006[3] sparked a debate in legal circles about the state of international law, and about the threats and opportunities arising from fragmentation. The term 'fragmentation' led to controversy chiefly because it seems to contain a negative bias as opposed to terms such as 'diversity'. Yet it is not necessary to frame the term negatively, in the same way that 'diversity' does not always have to lead to positive outcomes.

Let us first look into the common meaning associated with 'fragmentation'. As a starting point, Oxford Dictionaries define the term as follows: 'the process or state of breaking or being broken into fragments'.[4]

Three observations can be made at this stage. First, the very notions of 'fragments' or 'breaking' suggests that there once was, there still is or there will be something that is 'whole' or 'complete'. Second, fragmentation can be viewed as dynamic (a *process*) or static (a *state*). Third, the idea of fragmentat*ion* implies that there are driving forces behind it.

The first point has received particular attention in theoretical contributions on the fragmentation of international law.[5] The introduction of the term 'fragmentation' has forced authors to be explicit about their underlying assumptions (or made it easier for others to identify their implicit assumptions) on the nature of international law, and has brought scholarly disagreements to the foreground. This has led to a discussion about the extent to which international law can be seen as a 'system', as 'united', or whether there is an international legal 'order'. This discussion is not just a matter of semantics. Martti Koskenniemi and Päivi Leino were among the first to point to a possible political agenda behind the use of the term by several judges of the International Court of Justice at the turn of this century. They explained the judges' 'postmodern anxieties' concerning the unity of international law as an attempt to raise the profile of the International Court of Justice on the international plane at a time when an increasing number of other judicial bodies might undermine its relevance.[6] Likewise, the fears that some authors have expressed about the growing specialization in international law can be explained as a counter-reaction of general international lawyers afraid of

---

3 ILC 2006.

4 http://oxforddictionaries.com/definition/english/fragmentation?q=fragmentation (accessed 11 December 2013).

5 Craven 2005; Fischer-Lescano and Teubner 2004; Prost 2008; Martineau 2009; Prost 2012.

6 Koskenniemi and Leino 2002, 576–577.

becoming irrelevant within their profession.[7] Conversely, claims that international law is not (becoming) fragmented are also not entirely innocuous, and might well be understood as (sometimes disguised) arguments in defence of certain specialized regimes or judicial bodies.[8]

The real question underlying the debate on the term 'fragmentation' thus emerges: to what extent is there, or should there be, some kind of 'unity' in international law? Like 'fragmentation', the notion of 'unity' can also be interpreted in different ways.[9] The more contentious interpretation of 'the whole' would be an ideal that once existed or that should exist. This quickly leads to a loaded discussion of what such an ideal situation was or should be. Under a more neutral interpretation, however, 'the whole' could also simply be seen as the sum of its (fragmented) parts, irrespective of whether that is an ideal situation.

The second point is related to the first: does fragmentation refer to a *state*, i.e. the whole is (still) fragmented; or a *process*, i.e. what once was whole is now fragmenting?[10] The former view allows for a descriptive way of analysing fragmentation, whereas the latter forces one to be explicit about the starting and end points of the process (are we going 'up' or 'down'?) Either way, one is again faced with the need to define the 'whole' that is juxtaposed by its fragments.

The third point draws attention to the forces driving fragmentation. An initial response would be that fragmentation is the result of the proliferation of international institutions over time.[11] This proliferation itself is part of a broader trend: that of a move from territorial differentiation (based on national boundaries) to sectoral differentiation (based on the boundaries of specific issue areas).[12] This sectoral differentiation can in turn be seen as an expression of the emergence of different rationalities or world views emerging in modern societies,[13] or as legal pluralists Andreas Fischer-Lescano and Gunther Teubner put it: 'Societal fragmentation impacts upon law in a manner such that the political regulation of differentiated societal spheres requires the parcelling out of issue-specific policy-arenas, which, for their part, juridify themselves.'[14] Similarly, the ILC Study Group on fragmentation observes that '[t]he emergence of

---

7 Prost 2008, 158.
8 Martineau 2009.
9 Prost 2008; Prost 2012.
10 Martineau 2009, 4.
11 Raustiala 2013.
12 Leebron 2002, 7; Fischer-Lescano and Teubner 2004, 1009.
13 Khrebtukova 2008, 61–62.
14 Fischer-Lescano and Teubner 2004, 1009.

new "branches" of the law, novel types of treaties or clusters of treaties is a feature of the social complexity of a globalizing world'.[15] Furthermore, one of the key explanations for the existence of a fragmented international legal order is the absence of a centralized legislator, leading to a situation where 'there are as many law-makers as there are subjects of international law'.[16] Moreover, while it is sometimes assumed that states behave as unitary actors when involved in international negotiations, this assumption does not necessarily hold true.[17] It is therefore difficult to point to *the* will of a state when different ministries, as well as different people, with different mindsets are involved in negotiations in different forums.

These three preliminary points show that the use and definition of the concept of 'fragmentation' presents significant challenges. First, the notions of 'fragmentation' and 'unity' can be used as rhetorical devices to pursue a certain agenda. Yet likewise, other terms used in the literature such as 'diversity', 'pluralism' and 'polycentricity' have a positive subtext that would make them equally suitable to defend a certain political position with respect to the state of international law and governance.[18] Although using the term 'fragmentation' presupposes a 'whole' that once was, is, or will be, this whole does not need to be some form of absolute unity, comparable to domestic legal systems, where the normative hierarchy is clear, and the relationship between different laws can be determined on the basis of 'higher' laws. As I have argued before with colleagues, I 'assume neither an *a priori* existing state of universal order nor a universal trend towards order'.[19] This does not mean that I discard the idea of improving coherence of international law and global governance. As we will see in Chapter 5, there are various ways in which the consequences of fragmentation can be managed, and coherence could be enhanced; and there are indeed good reasons to do so, as explained in

---

[15] ILC 2006, para. 222.

[16] Pauwelyn 2006, para. 7.

[17] Chambers 2008, 50–51.

[18] Martineau 2009, 27. For instance, 'diversity' was contrasted with 'cacophony' in a special issue of the *Michigan Journal of International Law* focusing on the advantages and drawbacks of the fragmentation of international law. See, e.g., Simma 2004, 845. Pluralism is generally seen as a benign development by legal pluralists, who focus on the existence of overlapping legal systems. See, e.g., Fischer-Lescano and Teubner 2004. Polycentricity refers to 'many centers of decision making that are formally independent of each other': V. Ostrom *et al.* 1961, 831. For a discussion of polycentricity in a positive light, see E. Ostrom 2010.

[19] Biermann *et al.* 2009, 18.

Chapter 1. Second, a definition needs to clarify whether fragmentation is viewed as static (a snapshot) or dynamic (an ongoing process). Third, when using the term, it is important to have a sound understanding of the possible forces driving fragmentation.

For the purposes of this study, the term 'fragmentation' is used primarily as a descriptive device, referring to a state of fragmentation rather than a process. With fragmentation, I thus broadly refer to: the increased specialization and diversification in international institutions, including the overlap of substantive rules and jurisdictions. With this definition in mind, the next section examines various types of fragmentation.

3. THE MANY FACES OF FRAGMENTATION

The lack of conceptual clarity adds another layer of complexity to the debate on fragmentation. 'Fragmentation' means different things to different people and it is possible to distinguish a broad variety of 'fragmentations' in the literature.[20] By explaining different dimensions of fragmentation, I attempt to clarify some of the conceptual confusion.

A first distinction can be made between *substantive* and *institutional* fragmentation.[21] The ILC notably made this distinction when it decided not to examine 'the competence of various institutions applying international legal rules and their hierarchical relations *inter se*' (i.e., institutional fragmentation), instead focusing on 'the splitting up of the law into highly specialized "boxes" that claim relative autonomy from each other and from the general law' (i.e., substantive fragmentation).[22]

Despite the seemingly clear demarcation made in the ILC study between substantive and institutional fragmentation, the two types are very much interrelated. Georges Abi-Saab deems this a 'law of legal physics': 'To each level of normative density, there corresponds a level of institutional density necessary to sustain the norms'.[23] This relationship holds true in practice. For instance, the swordfish dispute between the EU and Chile is mostly seen as an example of institutional fragmentation, as the dispute was simultaneously brought before the WTO's dispute

---

[20] Cohen 2012.

[21] The use of the word 'institutional' in this specific context is different from the common usage in the international relations literature, and refers to international judicial bodies.

[22] ILC 2006, para. 13.

[23] Abi-Saab 1999, 925.

settlement mechanism and the International Tribunal for the Law of the Sea.[24] However, both dispute settlement mechanisms are inherently connected to substantive bodies of law, namely the various WTO agreements and the United Nations Convention on the Law of the Sea (UNCLOS).[25] There is hence a correlation between the extent to which norms and authority are fragmented: 'an increase in normative [i.e., substantive] integration generally results in a corresponding increase in authority [i.e., institutional] integration, and vice versa'.[26]

Second, fragmentation can be further divided into fragmentation along the lines of issue areas and fragmentation along geographical boundaries.[27] An example of the former is the debate on trade versus environmental regimes, whereas an example of the latter is the discussion on global versus regional trade and investment agreements.

Third, fragmentation may refer to the relationship between different interpretations of general international law, the relationship between general international law and specialized regimes, or the relationships among two or more overlapping specialized regimes.[28] An example of the first category is the *Tadić* case, in which the International Criminal Tribunal for the Former Yugoslavia came to a different judgment about the criterion to assess when an armed military group can be said to be acting on behalf of a foreign power than the International Court of Justice had before.[29] Under the second category, an example is the examination of how the general law of state responsibility relates to non-compliance mechanisms used in international environmental law or other more specialized regimes that may conflict with or complement the general rules.[30] This discussion pre-dates the ILC fragmentation study, and can be traced back to 1923, when the Permanent Court of International Justice (the predecessor of the International Court of Justice) in the *S.S. Wimbledon* case suggested that a specific body of rules may be 'self-contained'.[31] This notion of self-containedness re-emerged in 1980, when the International Court of Justice in the *Tehran-Hostages* case found that

---

24 See Orellana 2002.
25 Broude 2008, 182–183.
26 Broude 2008, 175.
27 Pauwelyn 2006, para. 3.
28 ILC 2006, para. 47.
29 *Prosecutor v. Dusko Tadić.*
30 Koskenniemi 1992.
31 *Case of the S.S. 'Wimbledon'*, 24.

the rules of diplomatic law constituted a 'self-contained regime',[32] raising the question whether specialized bodies of law could exist outside of general international law. This question drew particular attention in the cases of world trade law and EU law. Various authors have shown that the notion does not imply that a regime can be viewed in complete isolation from general international law, and that rather than speaking of 'self-contained' regimes it is better to speak of 'special' regimes.[33] The third category is exemplified by the various trade and environment disputes, of which the shrimp-turtle cases before the WTO Dispute Settlement Body are perhaps the most well known (see Chapter 8).

Fourth, one can differentiate between the fragmentation of primary and that of secondary norms. This binary distinction can be traced back to H.L.A. Hart's seminal book, *The Concept of Law*.[34] Whereas primary norms concern the principal rules of obligation, secondary rules can be regarded as rules about rules, governing the creation, interpretation and enforcement of primary norms.[35] The increase of primary norms after 1945 was an initial cause for concern, but attention has shifted towards the increasing number of secondary norms in support of these primary norms.[36] While secondary norms of special systems of international law may bring more order within those systems, they may also create more disorder in the overarching legal system.[37] For instance, norms determining when a state is in non-compliance with its obligations under the Kyoto Protocol are needed for the functioning of the treaty, yet at the same time they create uncertainty about the relationship between the Kyoto compliance mechanism and general norms of state responsibility.[38] More specifically, could a state that is not found to be in non-compliance with its emission reduction commitments under the treaty still be held responsible for any climate change damage that occurred?[39] Although the jury is still out on this question, there are strong arguments that general international law provides a 'fall-back' option in such cases.[40]

---

32 *United States Diplomatic and Consular Staff in Tehran (United States of America v. Iran)*, 38; see Simma 1985.
33 Lindroos and Mehling 2005; Simma and Pulkowski 2006.
34 Hart 1994, 89–109
35 Hart 1994, 94.
36 Marschik 1998, 212–213.
37 Marschik 1998, 239.
38 Verheyen 2005, 138–145.
39 Cf. ILC 2006, para. 190.
40 ILC 2006, para. 190; Verheyen 2005, 138–144.

Finally, one can speak more broadly of the fragmentation of sites of governance. Both international law and international relations scholars have pointed out that the state is no longer the only source for rules steering behaviour, and that a wide range of non-state actors, such as civil society organizations and businesses, play an increasingly important role. Political scientists have pointed to a shift from 'government to governance' at the domestic level,[41] and explored how hierarchical government intervention coexists with new forms of governance (e.g., self-regulation, voluntary agreements, codes of conduct).[42] At the international level, the term 'global governance' has been suggested 'as an analytical concept that provides a specific perspective on world politics different from that of "inter-national" relations'.[43] Some scholars further divide the concept of governance into public and private governance, and then refer to 'hybrid' forms of governance as the middle ground, involving both state and non-state actors.[44] What is important is that most analysts focus on norms and rules guiding social behaviour without requiring that those norms are generated by state actors.

International lawyers have also sought to grapple with the phenomenon of governance beyond the state. First, legal scholars have introduced the notion of 'global administrative law', acknowledging 'the vast increase in the reach and forms of transgovernmental regulation and administration designed to address the consequences of globalized interdependence'.[45] Specifically, the literature on global administrative law has investigated possibilities and challenges for addressing the 'accountability deficit' of these new forms of governance at the international level, through either the 'extension of domestic administrative law to intergovernmental regulatory decisions' or through 'the development of new mechanisms of administrative law at the global level to address decisions and rules made within the intergovernmental regimes'.[46] Second, scholars have shown an increasing interest in 'understanding, framing and taming the exercise of international public authority in the post-national constellation'[47] by examining the possibility to extend international institutional law to treaty bodies and informal regimes. Third, under the heading of 'informal

---

41 Rosenau and Czempiel 1992.

42 Jordan *et al.* 2005.

43 Dingwerth and Pattberg 2006: 198; see also Commission on Global Governance 1995; Rosenau 1995.

44 Bäckstrand 2008.

45 Kingsbury *et al.* 2005, 16.

46 Kingsbury *et al.* 2005, 16.

47 von Bogdandy *et al.* 2010, 26.

international lawmaking', scholars have sought to enhance understanding of the fact that 'the normative output that is the result of international (or transnational) cooperation does not fit our traditional understanding of international law-making as other instruments or procedures are used or different (non-state) actors take the lead'.[48] Fourth, the concepts of 'legal pluralism' and 'global legal pluralism' acknowledge that (international) lawmaking activities do not always involve the state.[49] Indeed, according to one advocate of global legal pluralism, 'it is now clear that the global legal system is an interlocking web of jurisdictional assertions by state, international, and non-state normative communities'.[50]

Under the guise of these various concepts ('global governance', 'international public authority', '(global) legal pluralism', etc.), scholars have sought to make sense of the variety of ways in which norms and rules are generated, how they influence social behaviour, and how legitimate and accountable they are. This dimension of fragmentation thus goes above and beyond the distinctions introduced above by acknowledging that behavioural change can be induced by norms that cannot be captured under traditional conceptions of 'law' or 'international law'.[51]

4. THE PROMISES AND PITFALLS OF FRAGMENTATION

While the very notion of 'fragmentation' may already reveal assumptions about its consequences, there is a large body of literature that provides arguments for and against fragmentation.[52] This section draws in particular on the more theoretical claims about the promises and pitfalls of fragmentation put forward in international legal discussions, although it will also refer to other studies that have discussed the consequences of fragmentation in the specific context of global climate governance.[53]

The ILC feasibility study on the fragmentation of international law conducted by Gerhard Hafner in 2000 indicated that the issue was one

---

48 Wessel 2011, 257; see also Wessel *et al.* 2011; Pauwelyn 2012.

49 E.g., Burke-White 2004; Fischer-Lescano and Teubner 2004; Berman 2007; Krisch 2010.

50 Berman 2007, 1159.

51 For a critical note on the 'governance mindset' from an international lawyer's perspective, see Koskenniemi 2004.

52 This section draws on van Asselt 2012a.

53 E.g., Biermann *et al.* 2009; Keohane and Victor 2011.

that should be looked at mainly in terms of 'risks', 'threats' or other negative connotations.[54] In particular, he argued that fragmentation can be seen as detrimental, since:

> [d]oubts could be raised as to whether international law will be able to achieve one of its primary objectives, dispute avoidance and the stabilisation of international relations and, thus, achieve its genuine function of law. The credibility, reliability and, consequently, authority of international law would be impaired.[55]

This rather general statement can be split up into various arguments against fragmentation. One argument often used is that the growing body of international legal rules threatens the unity and coherence of international law, as various specialized rules are created which allow international judicial institutions to come to diverging decisions (the institutional fragmentation referred to above).[56] For instance, a dispute between Ireland and the United Kingdom regarding the construction of a MOX plant reprocessing nuclear fuel led to three different legal procedures, each based on a different body of substantive law. The facts of the case allowed participants to the dispute to simultaneously frame it as an issue of the law of the sea, as a dispute between two EU Member States, and as one concerning the potential pollution of the North Sea.[57]

Another drawback is that fragmentation can arguably be used by a handful of powerful states to their advantage.[58] These states have the flexibility to opt for a mechanism that best serves their interests, and can create new agreements if the old ones no longer serve their interests.[59] With regard to dispute settlement, this may lead to 'forum shopping': countries are likely to choose the forum that is most likely to deliver a positive outcome. This explains why, in the above-mentioned swordfish dispute, the EU initiated proceedings at the WTO, arguing that Chile had restricted the movement of goods, while conversely Chile, the state taking conservation measures with respect to swordfish, brought its case before the International Tribunal for the Law of the Sea, alleging that the EU had violated the law of the sea. The swordfish dispute also shows that forum shopping is not only a strategy available for powerful states. Similarly, Helfer has argued that in the case of the regulation of

---

54 Hafner 2000.
55 Hafner 2000, 147.
56 Dupuy 1999; Kingsbury 1999.
57 ILC 2006, paras. 10, 439–442.
58 Benvenisti and Downs 2007; Drezner 2009.
59 Benvenisti and Downs 2007, 628.

intellectual property rights the states considered to be 'weaker' (i.e., developing countries, supported by non-governmental organizations) utilized a 'regime shifting' strategy to regulate intellectual property issues not through the WTO Agreement on Trade-Related Aspects of Intellectual Property Rights (TRIPS Agreement), but rather through the more favourable CBD, World Health Organization and the Food and Agriculture Organization of the United Nations.[60] In response, however, powerful states have engaged in regime-shifting strategies themselves, using bilateral and regional trade agreements to their advantage.[61]

Finally, a fragmented international legal system could lead to prioritization of certain fields of international law over others, for example, the dominance of international economic law over international environmental law.[62] Indeed, such prioritization may be inevitable, as 'each legal regime will naturally assert itself as the proper forum in which to address the situation, claiming superior status for its particular descriptions and concerns'.[63] Regimes, as Koskenniemi describes it, have a 'structural bias' in favour of themselves.[64] This structural bias becomes important when one regime can be considered 'stronger' because of the involvement of more powerful states or the presence of stronger compliance and enforcement mechanisms. This fear is often raised in the context of the trade and environment debate, where the WTO's dispute settlement system is generally considered to be stronger than the non-compliance mechanisms of most multilateral environmental agreements (MEAs).[65]

While plenty of arguments thus draw attention to the negative effects of fragmentation, it may also entail certain benefits. Indeed, after initial fears were expressed about the phenomenon, international legal scholars quickly realized that fragmentation might actually not be a problem but could also be seen as 'a kind of benevolent jigsaw'.[66]

First, fragmentation can be viewed as a positive indicator of increased diversity in legal norms and the expansion of international law to previously unregulated fields.[67] Over time, international law has come to cover important new issue areas of international relations such as international commerce, human rights and the environment. In other

---

60 Helfer 2004, 53–63.
61 Helfer 2009, 41.
62 Craven 2005, 5; ILC 2006, para. 493.
63 Khrebtukova 2008, 56.
64 Koskenniemi 2005, 600–615; Khrebtukova 2008, 67–71.
65 For a comparison, see González-Calatayud and Marceau 2002.
66 S. Humphreys 2012, 181.
67 Lindroos and Mehling 2005, 859.

words, '[s]pecial regimes and new organs are parts of an attempt to advance beyond the political present that in one way or another has been revealed unsatisfactory'.[68] However, while the expansion to new areas could in principle be seen as a positive development, this does not necessarily mean that 'more (international) law equals better (international) law'.[69]

Second, increased specialization is arguably a way of accommodating states' diverging interests. According to this view, states find that specialized regimes better serve their interests, thereby providing stronger incentives to comply. As Gerhard Hafner argues, a 'less-than-global approach seems particularly necessary when different States clearly hold different beliefs about what basic values should be preserved by international regulation'.[70] This argument resonates in the context of global climate governance, and is reflected in the calls for a 'minilateral' approach to international decision making on climate change (see Chapter 6).

Third, some lawyers have posited that fragmentation would not endanger the coherence of the wider body of international law, as it would lead to the global diffusion of the 'best ideas'.[71] Similarly, it has been argued by political scientists that regulatory competition may allow for the development of different solutions in different regulatory contexts, of which the most effective will 'survive' and be diffused to other regulatory contexts.[72]

5. CONCLUSION

Although the use of the term 'fragmentation' may serve particular political agendas, my modest claim is that it provides an accurate description of the current state of international affairs, where the emergence of different social rationalities at the global level has led to multiple regimes that overlap in terms of their subject matter. Whether the phenomenon is benign or malign is essentially in the eye of the beholder and further depends on whether the term is used to describe the relationship between different specialized regimes or the relationship between such special regimes and general international law. This book

---

68 Koskenniemi and Leino 2002, 578.
69 Benvenisti and Downs 2007, 602.
70 Hafner 2004, 859.
71 Charney 1999, 700.
72 Biermann *et al.* 2009, 27.

uses the term in its former sense and, in contrast with the studies discussed in this section, argues that the consequences of fragmentation do not necessarily depend on the existence of various overlapping regimes as such, but rather on how their interrelationships are managed. In order to analyse how regimes are (and can be) related, the next chapter turns to the subject of regime interaction.

4. Types of regime interaction

1. INTRODUCTION

At the same time that fragmentation became a fashionable topic among international lawyers, international relations scholars started to acknowledge that regimes and institutions could not be studied in isolation. While traces of this realization date back to the 1970s,[1] various research projects starting in the late 1990s put the issue of regime interaction high on the research agenda, especially in the area of global environmental governance.[2] Rather than synthesizing this extensive body of research,[3] I argue in this chapter that the starting points of the debates in international law and international relations are principally the same, and are founded on the increasing number of specialized regimes.

In the context of international environmental law, Edith Brown Weiss has referred to this phenomenon as 'treaty congestion'.[4] In international relations studies, the proliferation of regimes and international organizations has similarly been suggested to lead to a 'density of international regimes'.[5] In this context, Kal Raustiala and David Victor introduced the notion of 'regime complexes', referring to 'an array of partially overlapping and non-hierarchical institutions governing a particular issue area',[6] whereas Sebastian Oberthür and Olav Schram Stokke refer to 'institutional complexes' as 'two or more distinctive institutions that

---

[1] Biermann *et al.* 2009, 16.

[2] These include the Institutional Dimensions on Global Environmental Change project (O.R. Young 2002; O.R. Young *et al.* 2008); the project on Institutional Interaction, carried out by various European research institutes (Oberthür and Gehring 2006); and the United Nations University-Institute for Advanced Studies project on interlinkages (Chambers 1998; Chambers 2001; Chambers 2008).

[3] See Oberthür and Gehring 2011.

[4] Brown Weiss 1993, 697; see also Hicks 1999; Anton 2012.

[5] O.R. Young 1996, 1.

[6] Raustiala and Victor 2004, 279; see also Alter and Meunier 2009; Drezner 2009; Keohane and Victor 2011.

interact in their governance of the same activity, or aspects of the same activity, usually in a non-hierarchical manner'.[7]

Although the starting points may be similar, international lawyers have largely refrained from dealing with institutional aspects of a problem that is inherently connected to political realities, including its causes, its non-legal consequences, and the political responses.[8] Conversely, most international relations scholars have not conducted in-depth analyses of legal techniques for dealing with institutional interactions, and have not extended their findings to the debate on the fragmentation of international law.[9]

This chapter therefore not only aims to provide insights into types of regime interactions, but also seeks to serve as a communicating platform between the two disciplines. I do not seek to impose a general, overarching methodology for studying interactions, however, as such an approach does not do justice to differences between the disciplines. Instead, I aim to provide a framework for discussing the different types of interactions with a view to relating these to efforts to manage them.

2. TYPOLOGIES OF INTERACTIONS

There is a dearth of classifications and typologies in the literature on interactions, which facilitates the analysis of interactions, but at the same time complicates the identification of a first-best approach for such an analysis.[10] In addition, there is a Babylonian confusion about which terms should be used, with the different terms, 'interactions',[11] 'inter-linkages',[12] 'interplay',[13] 'linkages',[14] and 'overlap',[15] all referring to connections between overlapping institutions or regimes. To avoid

---

[7] Oberthür and Stokke 2012, 3.

[8] Notable exceptions include Wolfrum and Matz 2003; Chambers 2008; M.A. Young 2011a; M.A. Young 2011b; and Dunoff 2012.

[9] But see Zelli 2011b.

[10] Selin and VanDeveer 2003, 14; O.R. Young 2008, 154.

[11] Oberthür and Gehring 2006a; Gehring and Oberthür 2009; Oberthür and Stokke 2011b.

[12] Chambers 1998; Chambers 2001; van Asselt *et al.* 2005; Chambers 2008 Zelli *et al.* 2012.

[13] Stokke 2001; O.R. Young 2002; O.R. Young *et al.* 2008.

[14] E. Haas 1980; O.R. Young 1996; Charnovitz 2002; Leebron 2002; Selin and VanDeveer 2003.

[15] Rosendal 2001.

misunderstandings, I will stick to the term regime 'interaction', denoting that one regime may influence other regimes.[16]

Oran Young was among the first scholars to identify different types of interactions. His typology included: (1) embedded institutions, where regimes are embedded in overarching institutional arrangements (e.g., issue-specific regimes rooted in the notion of state sovereignty); (2) nested institutions, where specific arrangements (limited in scope) are nested in broader institutional frameworks in the same issue area (e.g., the framework-protocol approach used widely in international environmental law); (3) clustered institutions, where several not directly related institutions are intentionally tied together (as has happened in the package deals that formed UNCLOS and the WTO); and (4) overlapping institutions, where different institutional arrangements affect each other in a mostly unintentional way (e.g., trade and environment agreements).[17] Stokke refined Young's typology by introducing the notions of: utilitarian interaction, referring to cases where one regime alters the costs and benefits of behaviour under another; normative interaction, where one regime may confirm or contradict the norms of another; and ideational interaction, where one regime may learn from another.[18]

Sebastian Oberthür and Thomas Gehring further elaborated on the existing typologies by introducing a framework that emphasized the causal mechanisms of interactions. They distinguish between: (1) *cognitive interaction*, meaning that one institution is influenced by information or ideas stemming from another institution (similar to Stokke's 'ideational' interaction); (2) *interaction through commitment*, or the influence of the normative commitments entered into under one institution on another institution (overlapping with Stokke's 'normative' and 'utilitarian' interactions); (3) *behavioural interaction*, referring to behavioural effects triggered by one institution which have an influence on the effectiveness of another; and (4) *impact-level interaction*, where the ultimate targets of institutions interact.[19]

Some scholars have added a focus on the driving forces underlying regime interactions. The distinction introduced by Young between political and functional linkages clarifies that some interactions are a 'fact of

---

[16] Cf. Gehring and Oberthür 2006b, 4.

[17] O.R. Young 1996, 2–6.

[18] Stokke 2001, 10.

[19] I do not examine impact-level interaction, which, due to the natural-scientific complexities, is 'diffuse and difficult to analyse'. Cf. Oberthür and Gehring 2006a, 44.

life' (i.e., two institutions are linked in a biogeophysical or socio-economic fashion), whereas others 'come into play when actors forge links between issues and institutions *intentionally* in the interests of pursuing individual or collective goals'.[20] Oberthür and Gehring highlight this intentionality mainly in their category of cognitive interaction, arguing that intentionality means that one institution draws the attention of another institution to a particular issue. They call this particular type of cognitive interaction 'request for assistance', as opposed to the 'policy model', where learning takes place without such a request.[21] However, intentionality is not necessarily limited to this category, and may also take place with respect to interactions through commitment and behavioural interactions.

Finally, Oberthür and Gehring distinguish between interactions that have synergistic, conflicting (disruptive), or neutral/indeterminate effects on the target institution.[22] This distinction based on the outcome of interactions is of particular relevance for this study and will be discussed in more detail in Section 4 below.

3. THE OBJECT OF INTERACTION: INSTITUTIONS, REGIMES AND SOFT LAW

The specific study object (i.e., 'what' interacts?) in the interaction literature has largely been determined by disciplinary interests. Whereas international relations scholars have mainly focused on how institutions and regimes affect each other's development and performance, international lawyers have primarily examined international legal instruments such as treaties. In this section, I first explain why the focus of this study is on the notion of 'regimes' rather than 'institutions'. I then argue why, in addition, it is useful to also explicitly consider the interactions between soft and hard law.

In line with the institutionalist school of thought in international relations, many studies focus on international institutions as the interacting unit. One well-known definition by Robert Keohane describes international institutions as 'persistent and connected sets of rules (formal and informal) that prescribe behavioral roles, constrain activity, and shape expectations'.[23] A key rationale behind the focus on international

---

20 O.R. Young 2002, 23 (emphasis added).
21 Gehring and Oberthür 2006b, 14.
22 Oberthür and Gehring 2006a, 46.
23 Keohane 1989, 3.

institutions is that they are intentionally created to bring about behavioural change.[24] However, depending on the specific definition adopted, international institutions may refer to formal organizations, norms, or social practices, making it difficult to pinpoint the exact object of study.[25] For this reason, Oberthür and Gehring limited their study to negotiated institutions.[26] More importantly, in a study like this one, which aims to bridge the gap between two disciplines, the term 'institution' alone may lead to confusion as it generally has a different meaning to international relations and international law scholars; for international lawyers, 'international institutional law' more narrowly refers to the law of international organizations.[27]

The concept of 'regime' has also been the object of interactions in studies by both international relations[28] and international law scholars,[29] but at least here there seems to be more terminological convergence. The concepts of regimes and institutions are overlapping, since regimes can be seen as issue area-specific institutions.[30] The most widely used definition of regimes in the international relations literature was introduced in the early 1980s by Stephen Krasner: 'International regimes are defined as principles, norms, rules, and decision-making procedures around which actor expectations converge in a given issue-area'.[31] The idea of a set of principles, norms, rules, and decision-making procedures in a specific issue area has resonated with international lawyers (though clearly not with all[32]) and it is notable that the ILC's report on fragmentation frequently uses the term.[33] A regime is not synonymous with a treaty (although it can be based on one), but also includes decision-making procedures and organizational arrangements that may be constituted by an intergovernmental arrangement. For the purposes of this book, I adopt the definition proposed by Margaret Young (which is in

24 Oberthür and Gehring 2006a, 23.
25 O.R. Young 2002, 5–6.
26 Oberthür and Gehring 2006a, 23.
27 E.g., Klabbers 2002, 2.
28 Stokke 2001; Oberthür and Stokke 2011b.
29 M.A. Young 2011b; M.A. Young 2012b.
30 Levy *et al.* 1995, 270–271.
31 Krasner 1983, 3.
32 E.g., Koskenniemi 2009, 406–407.
33 ILC 2006.

turn adapted from Krasner's consensus regime definition) for the selection of the interacting objects: 'regimes are sets of norms, decision-making procedures and organisations coalescing around functional issue-areas and dominated by particular modes of behaviour, assumption and biases'.[34]

Most studies on regime interactions (both in international relations and in international law) focus on traditional, negotiated treaty-based regimes, which generally constitute 'hard law'. However, interactions may also involve regimes that are not based on legally binding instruments; in other words, they may involve 'soft law'. Soft law comprises instruments that are non-legally binding or that may be non-enforceable.[35]

The concept of soft law has been the subject of a long-standing debate in the legal literature, which can be broadly divided into two schools of thought.[36] The first, legal positivist school postulates that there is a binary distinction between law and non-law. Rejecting the notion of 'more or less binding' law, it questions the existence of soft law as such, as well as the appropriateness of the term.[37] The second, constructivist school argues that the notion of law in modern international affairs is less straightforward, and rather finds that there is a normative continuum comprising an 'infinite variety'[38] of informal instruments available to governments. Gregory Shaffer and Mark Pollack point out that the difference between the schools of thought lies mainly in whether one takes an '*ex ante* negotiations' or an '*ex post* enforcement' perspective. The latter requires a binary distinction (e.g., when an adjudicator needs to decide whether an instrument qualifies as 'law'), whereas the former emphasizes the diversity of instrument choices available to nation states.[39] In the same vein, Joost Pauwelyn suggests distinguishing between 'law' and 'having legal effects', and maintains that while conceptually the positivist, 'bright line' school is correct (i.e., law needs to be clearly distinguished from non-law), non-law, or 'soft law', may well have legal effects.[40]

Several definitions can be used to identify whether a specific governance arrangement can be considered as soft law. Kenneth Abbott and

---

34 M.A. Young 2012a, 11.

35 Boyle 1999, 901–902.

36 Shaffer and Pollack 2010, 712–717; see also Pauwelyn 2012, 127–129.

37 Klabbers 1996, 181.

38 Baxter 1980, 566; see also Schachter 1977; Boyle 1999; Hillgenberg 1999; Chinkin 2000.

39 Shaffer and Pollack 2010, 715–716.

40 Pauwelyn 2012, 127–130.

Duncan Snidal argue that '[t]he realm of "soft law" begins once legal arrangements are weakened along one or more of the dimensions of obligation, precision, and delegation'.[41] Softer law displays less of these features, and is thus more voluntary and vague.[42] Under this definition, it is possible that a treaty, usually considered hard law, exhibits features that soften it (see also Chapter 9). Another definition of 'informal international lawmaking' is introduced by Ramses Wessel and colleagues, who define it as:

> Cross-border cooperation between public authorities, with or without the participation of private actors and/or international organizations, in a forum other than a traditional international organization … and/or as between actors other than traditional diplomatic actors (such as regulators or agencies) … and/or which does not result in a formal treaty or traditional source of international law …[43].

These definitions will be used in this book to examine whether a certain governance arrangement can be considered soft law.

Soft–hard law interactions have been largely overlooked in studies of interactions, which have primarily focused on interactions between 'negotiated sectoral legal systems'.[44] Broadly speaking, however, the relationship between soft and hard law has drawn significant attention in the literature, which has mainly focused on how soft and hard law can function as alternatives or complements[45] but more recently has also indicated how they may act as rivals.[46]

The first way in which soft and hard law may interact focuses on their sequencing. Scholars have illustrated how soft law can 'harden' over time, for instance by spurring the development of customary international law.[47] In international environmental law, soft law, such as non-binding declarations (e.g., the 1972 Stockholm Declaration or the 1992 Rio Declaration), is often seen as a useful step in the process of developing hard law in the form of custom or legally binding treaties.[48] Non-legally

---

41 Abbott and Snidal 2000, 421; see also Abbott *et al.* 2000, 401.

42 Abbott and Snidal 2000, 422.

43 Wessel *et al.* 2011, 8.

44 Gehring and Oberthür 2006b, 5; but see Skjærseth 2006; Skjærseth *et al.* 2006; Vihma 2009.

45 Shelton 2011, 61–66.

46 Shaffer and Pollack 2010; Shaffer and Pollack 2011; Pollack and Shaffer 2012.

47 Chinkin 2000, 32; Shelton 2011, 61–62.

48 Shelton 2011, 62–66.

binding documents or declarations can indeed be used by states or non-state actors in the negotiations of new legally binding instruments and may shape the contours of emerging hard law through a process of legalization.[49]

Second, scholars have shown how hard law may be strengthened by complementary soft law arrangements, with soft law filling the gaps.[50] For instance, treaties can be supplemented by non-binding guidelines or interpretations. From this point of view, soft law plays a vital role in the progressive development of international law.[51] Soft law can thus support hard law without being hindered by some of the perceived drawbacks of the latter. In this regard, scholars point out that non-legally binding agreements are generally associated with greater flexibility and are usually easier to negotiate than legally binding treaties.[52]

Third, Shaffer and Pollack argue that the relationship between soft and hard law may also be antagonistic.[53] Rather than 'hardening' itself, soft law may actually 'soften' hard law by providing for an alternative forum. For instance, developing countries unhappy with the hard law provisions of the TRIPS Agreement have sought recourse to various soft law instruments, thereby creating more flexibility in the TRIPS Agreement.[54] Shaffer and Pollack argue that such antagonistic interactions are especially likely to occur 'if distributive conflict over the terms of cooperation is ubiquitous … , and if a given issue is subject to multiple regimes in the ever-thickening web of international norms, rules, and institutions'.[55] More specifically, they hypothesize that:

> Where powerful states disagree on policy, we are likely to see hard and soft law work in opposition to each other. Powerful states are likely to engage in forum shopping in such situations, advancing their interests by pressing for the adoption of legal provisions, both hard and soft, in forums that are most favorable to their respective positions. These overlapping hard- and soft-law oriented regimes, in turn, may come into conflict, with the result that soft-law oriented regimes could lose some of their technocratic and flexible characteristics and hard-law oriented regimes could become somewhat less determinate.[56]

---

49 Abbott *et al.* 2000.
50 Chinkin 2000, 30–31.
51 Shaffer and Pollack 2011, 1157–1158.
52 Raustiala 2005, 591–592; see also Abbott and Snidal 2000, 423.
53 Shaffer and Pollack 2010, 708; Shaffer and Pollack 2011, 1148.
54 Shaffer and Pollack 2010, 774–781.
55 Shaffer and Pollack 2010, 748.
56 Shaffer and Pollack 2010, 767.

Global climate governance, with its ongoing burden-sharing disagreements between powerful states, and an increasing number of governance arrangements, is thus a prime candidate for such interactions (see Chapter 6).

4. CONSEQUENCES OF REGIME INTERACTIONS

One of the distinctive features of regime interactions mentioned in the previous section merits a more detailed discussion, namely the consequences of interactions. The distinction between conflicting, synergistic and neutral effects is almost deceptively simple. Yet, as this section shows, determining the effects of regime interactions is anything but straightforward.

4.1 Conflicts

Before examining how conflicts can be resolved, it first needs to be established whether a conflict actually exists. The legal literature is split on the issue, with some authors arguing for a 'narrow' definition, and others favouring a 'wide' definition.[57] The main challenge in defining conflicts for the purposes of this study is to capture the various types of contradictions between different regimes. Before returning to this challenge, it is useful to provide an overview of the definitions that have been proposed (or endorsed) by legal scholars.

According to an early definition suggested by Wilfred Jenks in 1953, a 'conflict in the strict sense of direct incompatibility arises only where a party to the two treaties cannot simultaneously comply with its obligations under both treaties'.[58] This test of 'impossible joint compliance' has been supported by some international lawyers,[59] but has increasingly come under fire.[60] In particular, critics argue that Jenks' focus on *obligations* only is too limited, and excludes incompatibilities between obligations and permissions. This would include the case in which a multilateral environmental agreement permits a measure that restricts international trade, whilst a trade agreement contains a specific obligation not to restrict trade. Joost Pauwelyn's treatise on conflict of norms in international law therefore includes a proposal to expand the definition to

57 Vranes 2009, 10–38.
58 Jenks 1953, 426.
59 Marceau 2001, 1082.
60 Pauwelyn 2003a, 166–175; Vranes 2009, 19–21.

include conflicts involving permissive norms.[61] In another major contribution on the topic, Erich Vranes also argues for a broader definition based on a discussion of legal theory.[62] His definition comprises 'incompatibilities between permissions and obligations, permissions and prohibitions, and obligations and prohibitions', and argues that there is a conflict if one of the norms 'is necessarily or potentially violated'.[63] Jenks already acknowledged that his narrow definition might not cover all the divergences and inconsistencies between treaties that may have negative effects.[64] While the wider definitions proposed by Pauwelyn and Vranes ensure that certain obvious conflicts are not 'defined away',[65] even their constructions of conflict are insufficient to cover the various kinds of incompatibilities between regimes.

Therefore, while I do not reject the notion of a (legal) conflict between norms, I argue in favour of an additional definition for 'policy conflicts' that captures incompatibilities between regimes which need not be resolved through establishing a hierarchy between them (i.e., one of the norms necessarily prevails).[66] The ILC's broad conceptualization of conflict 'as a situation where two rules or principles suggest different ways of dealing with a problem'[67] would appear better suited to cover these 'policy conflicts'.[68] However, this definition would be too comprehensive, as 'different ways of dealing with a problem' might also lead to mutually supportive and complementary outcomes. It should thus be added that these 'different ways' lead to contradictory outcomes.[69] This seems at least broad enough to cover conflicts between regimes with diverging objectives not covered by the legal definitions proposed by Pauwelyn and Vranes.

The next (methodological) question is how to identify when regime interactions (could) lead to conflicts. Based on the various studies on interactions, I propose the following indicators:

Incompatible norms. This means that a normative conflict can be established according to the legal definitions outlined by Pauwelyn and Vranes. This is perhaps the clearest indicator of regime conflict, although

61 Pauwelyn 2003a, 166–175.
62 Vranes 2009, 19–21.
63 Vranes 2009, 38.
64 Jenks 1953, 426.
65 Vranes 2009, 20.
66 Cf. Fischer-Lescano and Teubner 2004, 1003.
67 ILC 2006, para. 25.
68 ILC 2006, para. 24.
69 See also Zelli 2011b, 200.

such cases may be rather limited in practice if it is assumed that states generally do not enter into agreements that could conflict with each other (the 'presumption against conflict'[70]).

Diverging objectives. This means that, at first glance, the objectives of two regimes are deemed incompatible. The case of trade liberalization versus environmental protection is most often used in this context[71] although, as discussed below, it may be difficult to pinpoint the precise objective for a regime.

The use of different principles and concepts. Regimes may embrace or apply principles and concepts that can be considered incompatible.[72] For instance, the precautionary approach, spurring action to prevent environmental degradation in the face of scientific uncertainty, may be at odds with an approach emphasizing cost-effectiveness.

Opposing economic incentives. One regime may provide economic incentives for behaviour that undermines the effective implementation of another regime. It is especially in the instances where states have a wide margin of discretion that there may be a conflict in the implementation phase.[73] A case in point is the CDM, which provided perverse financial incentives resulting in the continued production of ozone-depleting fluorinated gases, thereby hampering the effective implementation of the Montreal Protocol.[74]

'Negative' diffusion and learning. Although one regime's experiences may offer lessons for another (see below), 'if the contexts differ markedly, the ready availability of particular solutions from another regime may … crowd out options that would have been more appropriate, thus undermining ability to solve the problems addressed'.[75] In other words, lessons may be learned that could undermine the effectiveness of a regime.

Besides distinguishing between normative conflicts and policy conflicts, a distinction can be made between actual and potential conflicts. Metaphorically speaking, potential conflicts can be compared to a dormant volcano, whereas an actual conflict becomes visible when the volcano erupts. For instance, whereas the interaction between the WTO and MEAs is seen as a classic case of regime conflict, such a conflict has not manifested itself, for example, in the form of a challenge of

---

70 Pauwelyn 2003a, 240–244.
71 Wolfrum and Matz 2003, 9.
72 Wolfrum and Matz 2003, 7–8.
73 Wolfrum and Matz 2003, 11.
74 Oberthür *et al.* 2011, 117–118.
75 Stokke 2001, 11; Gehring and Oberthür 2006a, 329.

trade-related measures before the WTO's dispute settlement mechanism (see Chapter 8).

In subsequent chapters, I will make clear to which type of conflict I am referring: conflicts in the narrow legal sense (normative conflict) or in a broader sense (policy conflicts), as well as actual and potential conflicts.

4.2 Synergies

While the definition of 'synergy' has not been the subject of much debate, it nevertheless presents a tremendous challenge. For Oberthür and Gehring, there is synergy if the 'policy direction' of one institution is supported by measures taken by another institution.[76] Kristin Rosendal takes a slightly different approach, referring to the situation where the aggregate effects of two institutions are larger than the sum of effects produced on their own.[77] Her approach is in line with the definition provided in the Oxford Dictionaries: 'the interaction or cooperation of two or more organizations, substances, or other agents to produce a combined effect greater than the sum of their separate effects'.[78] The term has a positive connotation, associated with enhancing the effectiveness of one or both interacting regimes. It is in this sense that I use the term.

However, the terms used to define synergy ('policy direction', 'aggregate effects', 'combined effects') lead us to another question: how to determine the effectiveness of one regime, let alone the aggregate effects of several regimes influencing each other? This inevitably brings us to the matter of (regime) effectiveness.[79]

In the best case, assessing regime effectiveness is a highly complex task in which one needs to relate the stated or implicit objectives of a regime to observed or anticipated effects. This challenge has intrigued scholars over the years, with some opting for a case study approach[80] and others applying quantitative methods (such as databases).[81] Simple conceptions of regime effectiveness examine the output level (e.g., the

---

[76] Oberthür and Gehring 2006a, 46.

[77] Rosendal 2001, 97.

[78] http://oxforddictionaries.com/definition/english/synergy?q=synergy (accessed 11 December 2013).

[79] See, e.g., Brown Weiss and Jacobson 1998; Victor *et al.* 1998; Mehling 2004; Miles *et al.* 2002; Chambers 2008; Mitchell 2008; Breitmeier *et al.* 2011; O.R. Young 2011.

[80] O.R. Young 1999b; Miles *et al.* 2002.

[81] E.g., Helm and Sprinz 2000; Breitmeier *et al.* 2011.

decisions taken by regime bodies),[82] whereas more ambitious notions of regime effectiveness seek to determine how the regime has fared compared to an ideal outcome (also described as the 'collective optimum').[83]

In the worst case, regime effectiveness is an empty vessel, and any effort trying to infer any unambiguous regime objective is in vain. Michael Mehling illustrates this point neatly with reference to the UNFCCC, whose objectives consist of different aims based on countervailing interests. As a result, 'effectiveness can become little more than a normative judgment, based on subjective and, accordingly, contingent perceptions of the importance of certain objectives and the costs of their achievement'.[84]

Notwithstanding this sobering observation, efforts have been undertaken to operationalize 'effectiveness' in the context of overlapping regimes. Oberthür and Gehring base their classification of interactions explicitly on the literature on regime effectiveness. Their interactions correspond to three levels according to which a regime's effectiveness can be measured: the norms generated by the institution (*output*), corresponding to cognitive interaction and interaction through commitment; the behavioural effects on relevant state and non-state actors (*outcome*), corresponding to behavioural interaction; and the effects on the ultimate target of governance (*impact*), corresponding to impact-level interaction.[85]

Indicators at the regime output level provide the most tangible way of assessing whether regime interactions (could) lead to synergies, as the causal relationship between the regime and behavioural changes and/or impacts on the ground will be difficult to establish. For the purpose of this study, the indicators to determine the existence of synergies are as follows:

Shared principles and concepts. Although, as noted above, regime objectives are not always clearly observable, regimes may apply the same principles or concepts. For instance, environmental regimes may embrace the same principles, such as the precautionary principle or that of common but differentiated responsibilities and capabilities, strengthening their internal coherence.

Common economic incentives. Different regimes could provide economic incentives promoting the same type of activities. In terms of

[82] E.g., O.R. Young 1999b.
[83] E.g., Helm and Sprinz 2000; Hovi *et al.* 2003.
[84] Mehling 2004, 182.
[85] Oberthür and Gehring 2006a, 34.

financial incentives, they could even share the same financial mechanism.[86] For instance, the Global Environment Facility services a range of environmental conventions, including the UNFCCC and the CBD. By implementing cross-cutting programmes, it could simultaneously serve various regimes.[87]

Streamlined monitoring and reporting obligations. From an efficiency perspective, synergies can be created if monitoring and reporting obligations under different regimes are streamlined.[88] This could reduce data collection requirements at the national level, and decrease the administrative burden on states in terms of submitting reports on overlapping issues to different environmental conventions.

Shared supporting measures. Linking the supporting provisions of different agreements (related to capacity building, scientific cooperation, education and awareness, technology transfer, etc.) may lead to enhanced (cost-) effectiveness.[89]

'Positive' diffusion and learning. The 'robustness' of an agreement (i.e., its capacity to adapt to changing circumstances) can be improved by learning from the experiences and practices of other regimes or by the ability of the other regime to raise public support for a common cause.[90] Regimes may learn from each other's experiences in the formation or implementation stages. Although this may be difficult to prove empirically, learning may be observed if regime processes are closely related in terms of timing, participation and scope.[91]

In theory, it is possible to view synergies in a narrow or broader sense in a similar fashion as the distinction introduced for conflicts above. In a narrow sense, two norms may have synergistic effects. For instance, one norm may oblige states to tackle one cause of biodiversity loss by designating protected areas, whereas another norm may direct states to address another driver by taking climate change mitigation measures. In such cases, it is possible to speak of normative reinforcement. It is also possible that two regimes overlap to such an extent in terms of their objectives, approaches, principles and norms that they exert a positive influence on each other's effectiveness as a whole. This would be the case, for instance, if states conclude a regional environmental protection

---

[86] Chambers 2008, 151–152.
[87] Mee *et al.* 2008, 808.
[88] Chambers 2008, 147–151.
[89] Chambers 2008, 143–145.
[90] Chambers 2008, 145–147.
[91] Stokke 2001, 22.

agreement aimed at implementing or strengthening an existing multilateral agreement.[92] However, given that normative reinforcement would also entail broader synergies, they are discussed in conjunction.

Furthermore, as in the case of conflicts, it is possible that the aggregate effects may already be positive (actual synergies), or that there is a potential for synergies that remains to be captured (potential synergies). In the following chapters, I will indicate which type of synergy I am referring to.

Finally, it is worth noting that while regime interactions may lead to both conflicts and synergies, a third outcome is possible: that there is no (discernible) effect at all. This category of outcomes that are neither detrimental nor beneficial is usually referred to as 'neutral'.[93]

5. CONCLUSION

The fragmented state of international law and governance makes interactions between institutions increasingly likely. There are various ways to classify interactions and this chapter has provided a non-exhaustive overview of different typologies of interactions. These typologies provide insights into several key distinctive features of interactions that could help determine how interactions could best be managed. Table 4.1 provides a summarizing overview of these features. Where possible, these features will be identified in the following chapters, and I will return to them in the synthesis of this book.

The chapter has also explained how the consequences of regime interactions could be conflicting or synergistic, and has provided an in-depth exploration of the notions of both 'conflict' and 'synergy'. I have argued that conflicts could be viewed from a narrow (norm-based) and a broad (policy-based) perspective. The former has been the focus of most international lawyers, whereas the latter better captures the wider variety of possible relationships between different regimes. In addition, the chapter has explained that conflicts and synergies may have manifested themselves ('actual' conflicts/ synergies) or that they may still need to materialize ('potential' conflicts/synergies).

92 Another example is the 'institutional nesting' introduced by O.R. Young 1996, 3–4.

93 Oberthür and Gehring 2006a, 46.

Table 4.1 Distinctive features of interactions

Distinctive feature	**Options**
Object	• Soft law • Hard law
Causal mechanism	• Cognitive interaction • Interaction through commitment • Behavioural interaction • Impact-level interaction
Intentionality	• Intentional • Unintentional
Consequence	• Conflict – *Normative/policy* – *Actual/potential* • Synergy – *Actual/potential* • Neutral

5. Management of regime interactions

1. INTRODUCTION

Once it has been established that two regimes are, or may be, in conflict, or when there are potential synergies to be captured, interaction management could address conflicts or enhance synergies. But *how* should this be achieved? And *who* should take on this challenge?

This chapter explores these questions, first, by highlighting available strategies for the management of interactions. It provides an overview of different types of interaction management, and then introduces the main distinction of legal and political approaches to interaction management. This is followed by a discussion of various actors that are, or may be, involved in interaction management. These include notably actors within the interacting regimes. However, under the heading of 'autonomous interaction management' the chapter discusses the role that state and non-state actors can play in managing regime interactions. Lastly, it contains an indication of how to determine the success of types of interaction management.[1]

2. TYPES OF INTERACTION MANAGEMENT

Interaction management has been defined by Sebastian Oberthür as 'deliberate efforts by any relevant actor, group of actors, in whatever form or forum to address and improve institutional interaction and its effects'.[2] This very broad definition hints at a wide variety of approaches for managing interactions.

First, Oberthür identifies four *levels* of management. The highest level, overarching institutional frameworks, involves management going beyond the individual interacting regimes. Joint interaction management, in contrast, involves both interacting regimes, whereas unilateral management takes place by one regime only. Finally, autonomous interaction

[1] The chapter is an updated and expanded version of van Asselt 2011b.
[2] Oberthür 2009, 373; see also Stokke 2001, 11.

management may take place by states and non-state actors outside of the relevant regimes.[3]

Second, Oberthür separates two *modes* of interaction management. The first mode, regulatory interaction management, 'focuses on prescribing, proscribing or permitting certain behaviour, ascribing regulatory authority, and if paired with sufficient authority, implementing and enforcing measures against opposition'.[4] Enabling interaction management rather seeks to promote learning and capacity building.[5]

Interaction management may happen before the interaction occurs (*ex ante*), or after it has taken place (*ex post*).[6] For example, the use of provisions in treaties to regulate the relationship with other treaties may precede an interaction, whereas the resolution of a conflict before an international court usually takes place *post factum*.

In addition to these classifications, I distinguish two main types of interaction management for the purposes of this study: legal and political approaches. These two types include an amalgam of strategies to manage regime interactions that correspond with the first three levels of interaction management identified by Oberthür (overarching institutional frameworks, joint management and unilateral management). Legal approaches comprise a toolbox of legal techniques through which conflicts between treaties can be avoided and resolved and synergies can possibly be enhanced. These techniques may be applied by judicial bodies deciding on cases involving overlapping agreements or by policymakers and negotiators in drafting or revising agreements. Political approaches include activities by the bodies that are part of the interacting regimes with a view to institutional coordination. While the distinction between legal and political approaches (or: legal techniques and institutional coordination) provides a clear indication of the level to which international law can be relevant in addressing interactions, the two categories are undeniably interrelated: attempts to address interactions through legal techniques will involve political processes and may have political ramifications; conversely, coordination efforts may have legal implications.

---

3 Oberthür 2009, 375–376.
4 Oberthür 2009, 377.
5 Oberthür 2009, 377–378.
6 Stokke and Oberthür 2011, 7.

3. LEGAL TECHNIQUES

Legal techniques are primarily aimed at dealing with conflicts between treaties, rather than capturing synergies, although some of the techniques discussed below could be used for both purposes. Within the legal techniques targeted explicitly at conflicts, one can distinguish between conflict *avoidance* and conflict *resolution* techniques.[7] Put simply, conflict avoidance tools have the result that there is no conflict in the narrow, legal sense described in Chapter 4, whereas conflict resolution tools determine which norm prevails in case there is such a conflict.

The ILC holds an optimistic view of the role that legal techniques can play in addressing conflicts:

> The very effort to canvass a coherent legal-professional technique on a fragmented world expresses the conviction that conflicts between specialised regimes may be overcome by law, even as the law may not go much further than require a willingness to listen to others, take their points of view into account and to find a reasoned resolution at the end.[8]

The extent to which this positive assessment applies to managing regime interactions in global climate governance will be discussed in subsequent chapters. This section first introduces the various legal techniques, starting with conflict avoidance tools (treaty changes and drafting and treaty interpretation), followed by conflict resolution techniques (conflict clauses and priority rules). Important rules regarding the use of these techniques can be found in the Vienna Convention on the Law of Treaties (VCLT). This 'treaty on treaties' is not exhaustive on the rules of conflict avoidance and resolution, but the relevant provisions of the Vienna Convention will be referred to.

3.1 Treaty Changes and Drafting

Perhaps the most straightforward route (on paper) towards enhancing the compatibility of two treaties is changing one or both treaties. This is possible, first of all, through amendment of a treaty. Amendment procedures normally differ for each treaty.[9] The Agreement Establishing the WTO, for instance, requires acceptance by all WTO members.[10] In

---

7 Cf. Pauwelyn 2003a.
8 ILC 2006, para. 487.
9 VCLT, Arts. 39–40.
10 Agreement Establishing the WTO, Art. X.2.

contrast, an amendment to the UNFCCC needs to be adopted by consensus and, failing that, by a three-fourths majority vote.[11] Amendments can be problematic as they may require negotiations as lengthy as (or lengthier than) the negotiations of the original treaty. Furthermore, if amendments are not agreed upon unanimously, they create different and complex legal relationships between those parties accepting the amendment, and between them and the parties that do not accept the amendment.[12]

States party to one of the treaties may also decide to conclude an *inter se* agreement modifying a treaty. The Vienna Convention allows such agreements if these are provided for in a treaty or if these do not affect other parties or undermine the treaty's objectives.[13] This modification does not apply to third parties, and would thus address the relationship to the extent that states are a party to both the modified treaty and the other treaty.[14]

Although not precisely a modification of a treaty, another (drastic) form of dealing with the relationship of two treaties is by ending one of them. This could happen in the form of withdrawal from a treaty by one or more states, meaning that the tensions between two treaties will cease to exist at least for these countries. Alternatively, a treaty may be terminated under special conditions. Generally, if there is consent by all parties, withdrawal or termination is possible.[15] A treaty may also be terminated by the conclusion of a later treaty between all the parties relating to the same subject matter.[16] Finally, it may be possible that a party suspends or terminates the treaty when another party is in breach of the treaty because of compliance with another treaty's provisions.[17] However, these options treat the symptoms rather than the underlying problems, as they ultimately lead to a bilateral approach to addressing multilateral conflicts.[18]

---

[11] UNFCCC, Art. 15.3.

[12] Aust 2007, 262.

[13] VCLT, Art. 41.1; see also Art. 58.

[14] Interestingly, while *inter se* agreements have been discussed as a potential *source* of conflict in legal scholarship (i.e., are such agreements in conflict with multilateral treaties?), they have only occasionally been discussed as an avenue for *managing* regime interactions. See, e.g., Pauwelyn 2003a, 299–326; ILC 2006, paras. 295–323.

[15] Aust 2007, 288.

[16] VCLT, Art. 59.1.

[17] VCLT, Art. 60.

[18] Wolfrum and Matz 2003, 151–152.

While treaty modification (or termination) is thus one of the tools available in the legal toolbox for managing interactions, like lawmaking it is essentially political in nature, as treaty changes will be preceded by a political process of (re-)negotiation. This is therefore a clear example in which the line between political and legal approaches to interaction management is blurred.

3.2 Treaty Interpretation

Treaty interpretation is a technique that judicial bodies as well as diplomats and government officials may apply to harmonize two norms that seem to be in conflict.[19] Interpretation cannot resolve 'genuine' conflicts (i.e., cases in which compliance with one norm leads to breach of another[20]) and will always be limited to providing meaning to a term in a treaty that is insufficiently clear.[21] This means that the provision in another treaty must in some way be related to the ambiguous provision. It also means that the interpreted rule cannot simply be replaced by another provision. The main rules on how to interpret treaties are found in Articles 31 and 32 of the VCLT.

Article 31 of the Vienna Convention provides basic interpretation rules, stipulating that the ordinary meaning, context, object and purpose of a treaty are to be taken into consideration.[22] These rules provide scope for harmonizing two seemingly incompatible treaties. For instance, using a textual interpretation, examining the ordinary meaning of provisions as well as their (treaty) context, could indicate how parties intended to ensure compatibility between different treaties when negotiations were concluded. However, this can only happen within the boundaries provided by the text of the treaty.[23] A teleological interpretation could take into account the broader objectives of a treaty, thereby accommodating the goals of another treaty.[24] In other words, even though a specific part of a treaty may seem to be incompatible with another treaty, their broad objectives may still align.

---

19 Aust 2007, 230; see also Wolfrum and Matz 2003, 133–147.
20 Pauwelyn 2003a, 272.
21 Pauwelyn 2003a, 245; Wolfrum and Matz 2003, 146.
22 VCLT, Art. 31.
23 Wolfrum and Matz 2003, 135; Voigt 2008a, 270–273.
24 Wolfrum and Matz 2003, 135–138; Voigt 2008a, 278–279.

The VCLT also offers more dynamic interpretation rules, specifying that interpretation should take into account: (a) any subsequent agreement between the parties on interpretation of the treaty; (b) any subsequent practice in the application of the treaty; and (c) 'any relevant rules of international law applicable in the relations between the parties'.[25]

The first part of this provision refers to a possible 'authentic interpretation' that parties may adopt. 'Any subsequent practice' by states could indicate a tacit agreement on how to interpret certain provisions. However, verifying the existence of such an agreement would be difficult, as 'the tacit agreement must also be documented by equally tacit concerted action in regard to the implementation of the agreement'.[26] Where they are explicitly or implicitly mandated to do so, decisions by treaty bodies (such as the Conferences of the Parties to MEAs) could possibly be regarded as 'subsequent practice'.[27] However, given the uncertain legal status of activities by treaty bodies, the role of treaty bodies is further discussed under the heading of 'institutional coordination' below.

The ILC has stressed in particular the relevance of Article 31.3(c) of the VCLT. This interpretative axiom is at the heart of what has been termed the 'principle of systemic integration'.[28] Although there is a certain harmonizing appeal to the provision, no such principle has yet been explicitly recognized under general international law, and it still lacks an authoritative formulation.[29] It might therefore be more prudent to consider systemic integration an aspiration or objective rather than a principle. Still, the notion finds some support in rules of treaty interpretation and also past case law.[30]

A key question in the context of Article 31.3(c) is whether the 'relevant rules of international law' must be in place at the time of the adoption of a new treaty or at the time of interpretation. The latter allows for a more 'evolutionary approach' to treaty interpretation, and is arguably appropriate when interpreting terms that are likely to evolve over time.[31]

---

[25] VCLT, Art. 31.3(a–c).

[26] Wolfrum and Matz 2003, 143.

[27] Churchill and Ulfstein 2000, 641.

[28] McLachlan 2005, 279; ILC 2006, paras. 410–480; see also Sands 1999; Broude 2008; Lindroos and Mehling 2008; Pavoni 2010; Samson 2011.

[29] Lindroos and Mehling 2008, 268.

[30] McLachlan 2005.

[31] Pauwelyn 2003a, 264–268; ILC 2006, para. 478.

3.3 Conflict Clauses

The starting point in resolving conflicts between treaties is finding out whether states have sought to regulate these through conflict (or savings) clauses.[32] The general purpose of these clauses is to clarify the relationship between treaties. They can be found either in the main body of a treaty, or in its preamble.[33] There is a wide variety of possible formulations of conflict clauses, the most important of which are discussed here.

First, conflict clauses may grant priority to (specific) existing or future treaties or both of these. This subordination possibility is (somewhat superfluously) affirmed by the VCLT, which states that '[w]hen a treaty specifies that it is subject to, or that it is not to be considered as incompatible with, an earlier or later treaty, the provisions of that other treaty prevail'.[34] Treaties stipulating priority for existing treaties usually frame this by stating that the new treaty does not affect existing ones.[35] An example of a treaty granting priority to future agreements is Article 3.1 of the Convention on the Non-Navigational Use of Watercourses, which allows states to enter into new agreements adjusting the provisions of that treaty.[36] Giving priority to all other treaties is rare, and is usually phrased in a way suggesting that different treaties do not affect each other.[37]

Second, conflict clauses may claim priority over (specific) existing or future treaties, or over all treaties. An example of the first case is Article 311.1 of UNCLOS, which claims priority of UNCLOS over specific earlier treaties on the law of the sea.[38] The same Convention also exemplifies the second case by indicating that future agreements modifying or suspending provisions of UNCLOS cannot undermine the central provisions of the treaty.[39] The third case is most famously expressed in the Charter of the United Nations, which claims priority over all other international legal instruments.[40]

---

32 Pauwelyn 2003a; 328–361; Sadat-Akhavi 2003, 85–97; Wolfrum and Matz 2003, 120–128; Borgen 2005, 584–587; ILC 2006, paras. 267–294; Aust 2007, 218–227; Matz-Lück 2008a; Axelrod 2011.

33 Matz-Lück 2008a, para. 7.

34 VCLT, Art. 30.2.

35 Sadat-Akhavi 2003, 92.

36 Convention on the Non-Navigational Use of Watercourses, Art. 3.1.

37 Sadat-Akhavi 2003, 94–95.

38 UNCLOS, Art. 311.1.

39 UNCLOS, Art. 311.3; see Wolfrum and Matz 2003, 127–128.

40 UN Charter, Art. 103; see also VCLT, Art. 30.1.

Third, conflict clauses may determine priority on the basis of a seemingly objective indicator. For instance, various human rights treaties contain a clause granting priority to the 'more favourable' treaty in terms of human rights protection.[41] Several environmental treaties contain similar provisions allowing for the treaty with the highest level of environmental protection to prevail.[42] More often than not, however, it is unclear what these provisions entail exactly.

A few caveats should be made about the use of conflict clauses. First, as conflict clauses are part of a treaty, they do not apply to non-parties. In case of identical memberships, the relationship should be clear. However, when there are more or fewer parties in one treaty, the claim of priority or subordination only applies to those states that are parties to both treaties.[43] Second, conflict clauses may also be unable to clarify the relationship between treaties if they are unclearly or ambiguously phrased, thus not enabling the identification of a prevailing treaty. Third, adopting a treaty that contravenes another's treaty conflict clause does not necessarily make the new treaty null and void.[44] Finally, the clauses are not often used in judicial practice. As Jacob Werksman notes pessimistically, they may even never be applied 'in the absence of a single, unifying dispute settlement system'.[45]

3.4 Priority Rules

If a normative conflict cannot be avoided by treaty changes or interpretation, or resolved by conflict clauses, international law provides various rules for conflict resolution. The status of these rules in international law is not entirely clear.[46] What is clear, though, is that they are in principle available for the resolution of norm conflicts, and that they reflect the 'holy trinity' of the contractual freedom of states, the principle of *pacta sunt servanda*[47] and the principle of *pacta tertiis nec nocent nec prosunt*.[48] It is beyond the scope of this chapter to describe and assess all

---

[41] Sadat-Akhavi 2003, 213–232.

[42] McCabe 2007, 446–447.

[43] Matz-Lück 2008a, para. 4.

[44] Matz-Lück 2008a, para. 15.

[45] Werksman 1999a, 2.

[46] Vranes 2009, 41–42.

[47] 'Every treaty in force is binding upon the parties to it and must be performed by them in good faith'. VCLT, Art. 26.

[48] 'A treaty does not create either obligations or rights for a third state without its consent'. VCLT, Art. 34; see Pauwelyn 2003a, 327–328.

conflict resolution rules in detail.[49] This section therefore focuses on some of the most well-known priority rules, which can broadly be described with reference to the Latin terms *lex superior*, *lex posterior* and *lex specialis*.

Lex superior

The maxim *lex superior derogat lex inferiori* refers to the priority of hierarchically superior norms. Application of this maxim presupposes that there is or can be a normative hierarchy in international law. However, it is generally assumed that, with the possible exception of Article 103 of the UN Charter and *jus cogens*, there is no hierarchy of norms in international law.[50]

Article 103 of the UN Charter states that '[i]n the event of a conflict between the obligations of the Members of the United Nations under the present Charter and their obligations under any other international agreement, their obligations under the present Charter shall prevail'.[51] As noted above, since Article 103 is part of a treaty, it can also be seen as a far-reaching conflict clause. Although the legal consequences for acts contravening this clause (i.e., adopting a treaty that contradicts the UN Charter) are not entirely clear, most scholars seem to agree that the UN Charter does not make the conflicting treaty null and void.[52]

The idea of *jus cogens*, or peremptory norms from which no derogation is possible, is well accepted in international law. The possible contents of *jus cogens*, however, are not.[53] The VCLT contains two provisions that reflect the concept. Article 53 states that '[a] treaty is void if, at the time of its conclusion, it conflicts with a peremptory norm of general international law'. Such a peremptory norm, it continues, 'is a norm accepted and recognized by the international community of States as a whole as a norm from which no derogation is permitted and which can be modified only by a subsequent norm of general international law having the same character'.[54] In addition, Article 64 stipulates that '[i]f a new peremptory norm of general international law emerges, any existing

---

49 Pauwelyn 2003a; Sadat-Akhavi 2003; ILC 2006; Vranes 2009.

50 Weiler and Paulus 1997.

51 UN Charter, Art. 103; see ILC 2006, paras. 328–360.

52 ILC 2006, paras. 333–334; Matz-Lück 2008a, para. 15.

53 Weiler and Paulus 1997, 559; ILC 2006, paras. 361–379; Shelton 2006, 297.

54 VCLT, Art. 53.

treaty which is in conflict with that norm becomes void and terminates'.[55] The fact that there is no list of peremptory norms and no agreement on the criteria for listing such norms means that the substantiation of *jus cogens* primarily takes place through state practice and international case law.[56] International norms generally have to fulfil strict criteria to qualify, but there is a possibility that some norms of international environmental law may constitute *jus cogens* in the future.[57]

Lex posterior

Another well-known maxim, *lex posterior derogat legi priori*, essentially means that the more recent norm prevails over the older norm. While not mentioned as such, the Vienna Convention has effectively codified the notion of *lex posterior* by stating that '[w]hen all the parties to the earlier treaty are parties also to the later treaty ..., the earlier treaty applies only to the extent that its provisions are compatible with those of the latter treaty'.[58] The notion is based on the idea that the will of states may change over time and that new treaties derogating from older ones reflect this 'new' will. This assumption is valid if the membership of both treaties is identical, for instance in the case of two bilateral treaties between the same states, or between two multilateral treaties with the same membership.[59] This means that the *lex posterior* rule can be applied to a limited number of cases of conflict only.[60]

Crucially, the maxim introduces a temporal question: *when* is a treaty concluded? The Vienna Convention does not state whether the *lex posterior* rule (or, for that matter, a conflict clause) refers to: the date when negotiations finish, the date of adoption of a treaty, the date of opening for signature, the date of entry into force, or the date of signature, ratification or accession by states, etc.[61] Determining this is particularly difficult if the point of time of adoption is different for different countries,[62] a situation not uncommon in international environmental law.

---

55 VCLT, Art. 64.
56 ILC 2006, paras. 374–376.
57 Kornicker Ulhmann 1998, 104.
58 VCLT, Art. 30.3.
59 Vranes 2009, 56.
60 Wolfrum and Matz 2003, 153–154.
61 Vierdag 1988.
62 Vranes 2009, 60–61.

Lex specialis

Lastly, the maxim *lex specialis derogat legi generali* refers to a priority rule of 'special' norms over general norms.[63] Unlike the priority of *jus cogens* and the *lex posterior* maxim, no explicit reference is made to *lex specialis* in the Vienna Convention, although its basis in domestic law can be traced back to Roman times.[64] The rule reflects the idea that the more special norm offers the 'closest, detailed, precise and strongest expression of State consent'[65] and is most appropriate for particular circumstances. The rule could refer to a special norm providing a more detailed elaboration or application of the general norm, in which case there is not really a conflict between norms.[66] Two norms could also apply to the same set of facts but point in different directions, without any clear hierarchical relationship. In this case, *lex specialis* can come into play as a conflict resolution technique.[67]

The key question here is which of the norms is more 'specific' or 'particular'. This may be straightforward in some cases. For example, a protocol or other agreement implementing a framework treaty should arguably prevail over the more general framework agreement.[68] However, such cases will be rare, given the usually close interdependence between these types of treaties (e.g., the UNFCCC and the Kyoto Protocol). They rather represent the first type of *lex specialis* mentioned above, where one norm applies or elaborates upon another. The second type poses a more difficult challenge: dealing with treaties (or norms in treaties) that do not appear to have a clear relationship to each other but still apply to the same set of facts. The 'speciality' or 'generality' will depend on the facts of the case and is likely to be in the eye of the beholder. As the ILC notes, '[t]hat assessment is dependent on and makes constant reference to evaluative judgements of what is central and what marginal to a case, what aspects of it should be singled out and what aspects may be glossed over'.[69] Breaking down by specific norm is helpful, however. This is, for instance, how Vranes argues that the provisions restricting trade in ozone-depleting substances in the Montreal Protocol are *lex specialis* vis-à-vis the more general trade rules in international trade law, whilst noting his conclusions do not hold for the Montreal Protocol as a

---

[63] Lindroos 2005.
[64] ILC 2006, paras. 59–61; Vranes 2009, 63–64.
[65] Pauwelyn 2003a, 388.
[66] ILC 2006, para. 56.
[67] ILC 2006, para. 57.
[68] Wolfrum and Matz 2003, 156.
[69] ILC 2006, para. 106.

whole.[70] However, even in this arguably clear-cut case, it depends on how one approaches the issue. International trade law also contains rather specific norms about regulating trade; if these are applicable in relation to the trade restriction in an environmental treaty, they may equally be regarded as *lex specialis*.[71]

4. INSTITUTIONAL COORDINATION

4.1 Types of Institutional Coordination

Beyond the legal techniques discussed above, interaction management can take place through coordination and cooperation between decision-making and administrative bodies of the interacting regimes. Such coordination may come very close to an application of legal techniques. For instance, COP decisions with relevance to another regime can be regarded as a form of treaty interpretation (see below).[72] The line between legal techniques and institutional coordination is even thinner in the case of treaty changes: the negotiation process leading up to such changes could be seen as a form of institutional coordination, whereas the outcome of such a process could be seen as state actors applying a legal technique.

'Coordination' could range from loosely coupled arrangements to tightly integrated systems.[73] Coordination is possible between (groups of) actors involved in the interacting regimes with a view to addressing conflicts and enhancing synergies without resorting to legal means (e.g., dispute settlement mechanisms). Institutional coordination can take many forms, but a few crude distinctions can be made:

Unilateral or joint. In line with Oberthür's distinction between unilateral and joint interaction management, it is possible to distinguish institutional coordination involving actors in only one of the interacting regimes, and joint coordination involving both regimes.[74] The former includes unilateral decisions adopted under one regime that affect the outcomes of the interactions, but comprises also other (administrative) actions within the institutional boundaries of one regime. The latter necessarily involves communication between the two interacting regimes.

---

70 Vranes 2009, 368–369.
71 See also Safrin 2002, 606.
72 Churchill and Ulfstein 2000, 641.
73 Jordan and Schout 2006, 40–44.
74 Oberthür 2009, 376.

Ad hoc or structural.[75] At times, coordination consists of a one-off political effort in response to current affairs or a decision in another regime. This could be limited to a unilateral acknowledgement of the changes in another regime, or could entail a response to it. It could also involve the exchange of information without any follow-up. Coordination could also be more structural, by establishing ongoing relationships between the interacting regimes, or by dealing with those interactions through an overarching structure. This could be achieved, for instance, by putting in place a consultation mechanism between the different regimes. Far-reaching forms of structural coordination include the establishment of a permanent body or organization to manage interactions. In international environmental law, the United Nations Environment Programme (UNEP) is such a permanent body that is mandated '[t]o further the development of its international environmental law aiming at sustainable development, *including the development of coherent interlinkages among existing international environmental conventions*'.[76] Even more structural forms of coordination have been proposed, such as clustering different treaties or creating a World Environment Organization (see Chapter 7).

Legal basis. This is important in two respects. First, there is the question of legal personality of treaty bodies, which would allow such bodies to enter into formal contractual arrangements specifying inter-institutional coordination.[77] For instance, treaty bodies could enter into a formal arrangement by concluding a Memorandum of Understanding or Memorandum of Cooperation.[78] However, they can also exchange information or communicate through informal channels or coordinate informally through observing meetings organized under the umbrella of another regime.[79] Second, treaties occasionally spell out the purpose of coordination efforts in detail; i.e., they provide a clear mandate for coordination. Other provisions, however, are more open-ended regarding the outcomes, and do not suggest a specific course of action.

---

75 See also Jordan and Schout 2006, 41–44.

76 UNEP 1997, para. 3(b) (emphasis added).

77 Chambers 2008, 60–66.

78 Wolfrum and Matz 2003, 173–174; Chambers 2008, 66–67; Scott 2011, 192–200.

79 Although observership is often based on a formal act (accreditation as an observer) by a regime's decision-making body.

Treaty bodies involved. Coordination may involve the decision-making bodies of regimes, such as the COPs of environmental treaties; administrative bodies, such as treaty secretariats; or judicial bodies, such as the dispute settlement bodies of the WTO. These actors are discussed in turn below.

4.2 Actors in Institutional Coordination

Decision-making bodies

Because of their intergovernmental nature, activities of decision-making bodies could have an important impact on the interaction. These bodies could adopt decisions addressing an existing or looming conflict, or enhance synergies with another regime. If a particular conflict is caused by the action of a decision-making body in the first place, another decision could seek to 'overrule' the first decision. But a decision could also be adopted to respond to policy developments in the other regime, possibly with a view to counteracting the effects of those developments.

As noted above, COP decisions can amount to treaty interpretation through a 'subsequent agreement' under Article 31.3(a) or 'subsequent practice' by parties under Article 31.3(b) of the VCLT. Through interpretation, COP decisions can clarify ambiguous provisions of the underlying treaties[80] with a view to harmonizing them.

Activities by the decision-making bodies could also be seen as international lawmaking. For instance, Jutta Brunnée suggests that '[a] distinctive feature of MEA-based law-making processes is that they take place under the *aegis* of the treaty's institutional core, the COP'.[81] Brunnée points out that, from a formal point of view, lawmaking by treaty bodies might be considered to be based on a form of implied state consent.[82] However, treaty body activities are hardly ever entirely autonomous. Expanding on Brunnée's work, Annecoos Wiersema concludes that 'consensus-based COP activity … cannot be seen as giving rise to stand-alone legal or even political obligations' and that COP decisions 'hold little meaning but for their connection to the treaty'.[83] It is thus evident that the legal force of COP decisions is intrinsically connected to the treaty obligation upon which they are based. The question of whether

---

[80] Churchill and Ulfstein 2000, 641; Brunnée 2002, 31.

[81] Brunnée 2002, 4; see also Churchill and Ulfstein 2000; Brunnée 2005; Wiersema 2009.

[82] Brunnée 2002, 23–31.

[83] Wiersema 2009, 245; see also Brunnée 2002, 24; Fitzmaurice and Elias 2005, 262.

COP decisions amount to lawmaking thus necessitates a case-by-case analysis.[84] The more important point, however, is that COP decisions are not devoid of normative substance,[85] and in practice may have a significant impact on the behaviour of state and non-state actors and may thereby influence the relationship between two regimes.

Decision-making bodies often delegate activities involving communications with another regime to its administrative bodies (see below). However, they may also be more directly involved, meaning that joint decision making is possible. For instance, in 2010 the decision-making bodies of three autonomous MEAs (the Basel Convention on the Control of Transboundary Movements of Hazardous Wastes and their Disposal, the Rotterdam Convention on the Prior Informed Consent Procedure for Certain Hazardous Chemicals and Pesticides in International Trade, and the Stockholm Convention on Persistent Organic Pollutants) convened in an extraordinary meeting, leading to a set of 'omnibus decisions' adopted under each of the treaties. The decisions endorsed a number of ways in which further synergies between the conventions could be pursued, primarily emphasizing the potential for efficiency improvements through, for example, the appointment of a joint head to the bureaucracies and the identification of joint services.[86]

Bureaucracies

Bureaucracies, such as the UNFCCC secretariat, are important actors in global environmental governance, yet they have largely flown under the radar of analysts.[87] Studies by international relations scholars show that bureaucracies can exert different types of influence: (1) cognitive influence, where bureaucracies affect the knowledge and belief systems of political actors; (2) normative influence, where bureaucracies affect the

---

[84] Röben 2000, 374.

[85] Following the *ex ante* approach to soft law, it would also be possible to classify COP decisions as soft law. However, this would simplify the role of COP decisions, and '[obscure] what COP activity is actually doing in relation to the underlying treaty obligation'. Wiersema 2009, 264.

[86] UNEP 2010, Annex I, paras. II.3 and III.3; see Scott 2011, 205–207.

[87] See the various contributions in Biermann and Siebenhüner 2009b. I differentiate between bureaucracies and international organizations. With bureaucracies, I refer to the administrative apparatus of a specific regime. Cf. Biermann *et al.* 2009, 39–40.

normative development of a regime; and (3) executive influence, where bureaucracies assist countries in implementing international agreements.[88]

Bureaucracies can be regarded as key actors 'behind the scenes' in managing interactions. Although their influence on the individual regimes they are tied with is becoming clearer, their role in managing the relationships between different regimes remains under-researched.[89] Bureaucracies may be able to exert the types of influence mentioned above also in managing interactions. With respect to cognitive influence, secretariats could raise awareness of interactions and their consequences. Regarding normative influence, secretariats could seek to influence rule development in one or both of the interacting regimes. In terms of executive influence, secretariats could help build capacity in countries to ensure that countries meet their commitments under the interacting regimes.

The involvement of bureaucracies in interaction management raises several legal questions. As noted above, a first question is whether secretariats have the legal capacity to enter into formal external cooperation agreements. This has been examined in depth by Robin Churchill and Geir Ulfstein, who conclude that MEA treaty bodies 'have implied powers to act on the external plane, including the capacity to enter into treaties when necessary to carry out their functions'.[90] Specifically with respect to secretariats, Bradnee Chambers argues that while the legal personality of secretariats may not be entirely clear, their power 'would certainly include entering into agreements of collaboration with other MEAs where there is a clear overlap or interest'.[91]

Depending on their specific mandates, bureaucracies can engage in a bilateral dialogue with other administrative bodies. This could take place through the formal method of Memoranda of Understanding/Cooperation, or informally. There is some debate about the legal nature of Memoranda of Understanding/Cooperation and other formal agreements between bureaucracies. Chambers, for example, argues that they are 'international agreements under international law'.[92] Rüdiger Wolfrum and Nele Matz, in contrast, note that they are 'not legally binding treaties but are only politically binding upon the participating

---

88 Biermann *et al.* 2009, 47–49.
89 Notable exceptions include Jinnah 2010; Jinnah 2011.
90 Churchill and Ulfstein 2000, 649.
91 Chambers 2008, 66.
92 Chambers 2008, 66–67.

institutions'.[93] Ulfstein occupies the middle ground, arguing that 'internationally binding agreements are not necessary or desirable for such coordination between secretariats'.[94] None of the authors is necessarily wrong: Memoranda of Understanding/Cooperation are unlikely to have the same legal nature as legally binding agreements between states; however, depending on their substance, they may well have binding effects on the bureaucracies.[95]

In some cases, specific forums have been established for MEA secretariats, allowing for a more structural discussion about regime overlaps. For instance, the CBD has been involved in the creation of two 'liaison groups', in which it participates alongside the secretariats of various other environmental treaties. A Biodiversity Liaison Group, which comprises the secretariats of six biodiversity-related conventions, was established in 2004, and seeks 'to enhance synergies and reduce inefficiencies', mainly at the national level.[96] The 2001 Joint Liaison Group, which will be discussed in more detail in Chapter 7, involves the secretariats of the three 'Rio Conventions' (the CBD, the UNFCCC and the UN Convention to Combat Desertification). Another example in the area of global environmental governance is the Collaborative Partnership on Forests, which seeks to foster collaboration between 14 organizations with 'substantial programmes on forests', including the secretariats of several regimes.[97]

Moreover, there has been activity at the United Nations level with a view to fostering coordination among UN bureaucracies. In the area of international environmental governance, an important coordination role has been assigned to UNEP.[98] In addition, various inter-agency coordination mechanisms have been established within the UN. For instance, climate change is one of the thematic areas of the UN Chief Executives Board for Coordination, which established a Climate Change Action Framework in 2011.[99] However, the UNFCCC Executive Secretary does not form a part of this group. More generally, as Chambers finds, 'MEA secretariats are isolated from the internal UN coordination mechanisms,

---

93 Wolfrum and Matz 2003, 173.
94 Ulfstein 2007, 888.
95 Scott 2011, 193.
96 CBD Decision VII/26, para. 1; see Caddell 2012.
97 Collaborative Partnership on Forests 2003.
98 Chambers 2008, 71–80.
99 http://unsceb.org/content/ceb-climate-change-action-framework (accessed 11 December 2013).

since there are no financial regulatory incentives'.[100] Rather than depending on overarching structures for coordination, MEA bureaucracies thus play an important role themselves.

Dispute settlement bodies

It may seem strange to mention dispute settlement bodies in a section on institutional coordination, given the earlier discussion of legal techniques which seem, at first sight, more applicable to these bodies. However, there is scope for institutional coordination involving dispute settlement bodies without resorting to the legal techniques discussed above. The aim of this coordination would normally be to enhance adjudicators' understanding of other regimes. This could happen in various ways, including consultations with experts. These could be scientific experts, experts from other international organizations, as well as non-governmental organizations or other actors (through *amicus curiae* briefs).[101] In addition to relying on external experts, it is also possible to require adjudicators to have specialized knowledge of other regimes (e.g., by appointing an international environmental law specialist in an investment arbitration).

5. AUTONOMOUS INTERACTION MANAGEMENT

Legal techniques and institutional coordination, as described above, cover most forms of 'collective interplay management',[102] i.e., interaction management that goes beyond implementation by individual actors, such as governments, civil society, businesses or public–private partnerships. However, choices made by individual states and non-state actors in response to regime interactions may be equally (or perhaps even more) important, and constitute a form of 'autonomous' interaction management. The difference between mere implementation of international agreements and autonomous interaction management is that the latter is undertaken with a view to addressing the relationship between two regimes.

The most straightforward rationale for autonomous interaction management would be enhancing the mutual supportiveness of two regimes. For a state, this rationale could be based on the notion of *pacta sunt servanda*: states are supposed to implement the international agreements

100 Chambers 2008, 78.

101 Pauwelyn 2002; M.A. Young 2011a, 206–224.

102 Oberthür 2009, 376.

they entered into in good faith. However, autonomous interaction management may also be undertaken with a view to confirming inconsistency. In this regard, Oberthür mentions the example of a state adopting a trade-related environmental measure with a view to provoking 'either tacit acceptance by other actors or a decision under the WTO dispute settlement procedure that explicitly permits such restrictions under WTO law'.[103]

Non-state actors may also be involved in autonomous interaction management. For instance, environmental non-governmental organizations that seek to reconcile the objectives of various environmental regimes may seek to raise awareness or implement projects on the ground that take into account aspects of the various regimes. Similarly, it has been argued that public–private partnerships involving both state and non-state actors can play a vital role in interaction management.[104]

6. THE EFFECTIVENESS OF INTERACTION MANAGEMENT

The different typologies facilitate an analysis of interaction management activities, but leave a very important question unanswered: when is interaction management *effective*? This question can be approached in various ways.

First, one could deem interaction management successful if it raises awareness and increases knowledge among relevant (state and non-state) actors about the issues of overlap between different regimes, and the consequences of regime interactions.[105] Although a shared understanding of interactions could more accurately be seen as a *condition* for effective interaction management,[106] it constitutes an important first-level indicator of the performance of interaction management.

Second, interaction management could be considered effective if it results in efficiency gains. These gains could be achieved by lowering the costs of participation in both regimes, for instance, if diverging obligations (e.g., reporting) or activities (e.g., scientific research) are streamlined.

Third, a more ambitious approach would relate interaction management to notions of regime effectiveness, arguing that if the effectiveness

---

[103] Oberthür 2009, 376; Stokke and Oberthür 2011, 10.
[104] Visseren-Hamakers *et al.* 2011.
[105] Oberthür and Stokke 2011a, 319–320.
[106] Oberthür and Stokke 2011a, 324.

of one of the regimes is enhanced, the management efforts are successful. It could also be argued that interaction management is successful if the effectiveness of *both* regimes is enhanced or, at the very least, the effectiveness of both regimes is not reduced.[107] However, evaluating how interaction management performs with reference to notions of regime effectiveness will face the same challenges I noted above for studies on regime effectiveness generally, compounded by the fact that it will be even more difficult to identify the causal mechanism by which interaction management has altered the consequences of regime interactions.

Fourth, the performance of interaction management efforts could be assessed by determining whether a normative conflict has been avoided or resolved, without any preference given to *how* this has happened. This could, for instance, be the case if a court applies one of the conflict resolution rules to a specific conflict, thereby indicating that one of the two interacting regimes prevails.

The fifth and most ambitious approach would be to relate interaction management to an overarching objective or, as Jeffrey Dunoff puts it, a 'redemptive narrative' for reconciling two regimes.[108] The concept of sustainable development has been suggested as such an overarching objective,[109] but other options (e.g., human rights, *jus cogens*, obligations *erga omnes*) or approaches (e.g., global constitutionalism[110]) are also conceivable. However, the identification of such narratives is challenging, as their contents remain elusive and contested, and the narratives can themselves become part of the struggle between regimes (see Chapter 10).[111]

What becomes clear is that operationalizing most of these alternatives is fraught with difficulties and will depend on subjective value judgements of what the outcomes should be of regime interactions. While acknowledging the inherent limitations and subjectivity of any overarching standard for determining the effectiveness of interaction management, the following pragmatic approach, which comprises several of the above-mentioned possibilities, will be applied in the subsequent chapters to determine the impact of interaction management. Interaction management in the case studies will be deemed (at least partially) effective if it can be shown: (1) that it fosters a shared understanding of the regime interactions and their consequences; (2) that it leads, or may lead, to

107 Cf. Levin *et al.* 2008.
108 Dunoff 2012, 155; see also Trachtman 2011, 16.
109 Voigt 2008a.
110 E.g., Dunoff and Trachtman 2009b; Klabbers *et al.* 2009.
111 Koskenniemi 2012, 321.

efficiency gains that reduce the burden of simultaneous implementation of both regimes; and/or (3) that a normative conflict has been avoided or resolved.

7. CONCLUSION

The last three chapters together have laid the groundwork for the following chapters by introducing the key concepts used in this book, explaining how and with which effects overlapping regimes can influence each other and showing a variety of ways in which such regime interactions can be managed.

This chapter has provided an overview of the various techniques that can be found in the legal toolbox, distinguishing between conflict avoidance and conflict resolution techniques. In addition, it has discussed various forms of institutional coordination, as well as the different types of actors that can be involved in such coordination. In addition to these forms of 'collective interaction management', the chapter has also briefly touched upon the notion of 'autonomous' interaction management, which does not involve the interacting regimes or overarching institutional structures.

The chapter has given an indication of the difficulties associated with providing an objective indication of the effectiveness of interaction management, explaining how the focus of this book will instead be on indications of shared understanding of regime interactions, efficiency gains and the avoidance or resolution of normative conflicts.

The following chapters will apply the analytical framework to three pertinent cases in global climate governance, focusing on interactions between the UN climate regime and: minilateral clean technology agreements (Chapter 6); the Convention on Biological Diversity (Chapter 7); and the World Trade Organization (Chapter 8). In each chapter, I will look specifically at the consequences (in terms of conflicts and/or synergies) of regime interactions, as well as at the potential for legal techniques, institutional coordination and autonomous interaction management to address actual or potential conflicts and to enhance synergies. I will return to the classifications introduced here in the synthesis in Chapter 9.

PART III

Case studies

6. The UN climate regime and minilateral clean technology agreements

1. INTRODUCTION

This chapter presents the first case study, analysing regime interactions within a narrowly defined, single issue area. More specifically, the chapter focuses on interactions between the multilateral UN climate regime and 'minilateral'[1] regimes in the same issue area, as exemplified by minilateral clean technology agreements.[2] The chapter seeks to provide insights into the consequences of interactions between the climate regime and such agreements, as well as ways in which interactions have been or can be managed. By doing so, it also aims to highlight how, through institutional coordination, the UN climate regime and minilateral clean technology agreements could come to an acceptable division of labour that makes use of their respective strengths. It thus seeks to show how minilateral initiatives and the climate regime could be complements rather than competitors.

One way of framing the climate change problem is by highlighting its technological dimensions. By adopting a technology-based perspective, the genesis of the problem can be traced back to the technological innovations that enabled industrialization, and thereby led to rapidly increasing greenhouse gas emissions. The solution can equally be framed in terms of technological change. In particular, promoting technological change through research and development (R&D) on, and the large-scale

---

[1] By 'minilateral', I mean initiatives with participation by three or more countries (i.e., excluding unilateral or bilateral initiatives), but whose membership is still limited on the basis of certain country characteristics (i.e., they are not open to every country).

[2] Cf. de Coninck 2009, 214. The term 'clean technology agreements' used here refers to agreements among governmental actors, with or without private sector participation, and adopted through an agreement that may or may not be legally binding.

diffusion and deployment of, low-emission technologies is seen as a crucial component of any successful climate policy. Which combination of clean technologies should be pursued, however, remains an open question.

The climate treaties contain various provisions to promote cooperation on clean technology development and diffusion. At the same time, several country coalitions have concluded clean technology agreements to promote technologies in which they have a particular interest. The co-existence of multilateral and minilateral approaches to promote clean technologies leads to the questions of how they relate to each other, and what can be done to make them mutually supportive with a view to achieving the common goal of reducing global greenhouse gas emissions.

This chapter examines the relationship between the multilateral UN climate regime and minilateral clean technology agreements, using the example of the now-defunct Asia–Pacific Partnership on Clean Development and Climate. The APP is chosen not only as a prime example of a minilateral clean technology agreement, but also as an interesting case of an agreement that was initially presented as an alternative to the Kyoto Protocol by a limited number of countries, was then downplayed as a more practical approach complementing the climate regime and eventually withered away into oblivion. Furthermore, the reasoning behind the adoption of the APP (which focused on the world's major emitters) forms an ongoing argument for the pursuit of a minilateral rather than a multilateral climate agreement.[3] While minilateral clean technology agreements like the APP may complement the UN climate regime, they also form part of a competing discourse that exhibits dissatisfaction with the approach the climate treaties, and in particular the Kyoto Protocol, have adopted so far.[4]

2. THE GOVERNANCE OF CLIMATE CHANGE AND CLEAN TECHNOLOGIES

2.1 Climate Change and Clean Technologies

Deploying low-emission technologies and improving existing technologies is of major importance in the battle against climate change.[5]

---

[3] Naím 2009; Hale 2011, 97; Victor 2011, 242–243.

[4] McGee and Taplin 2009b; McGee and Taplin 2012.

[5] While there are various definitions of 'technology', I use it here in a broad sense, including both the 'hardware' (equipment, infrastructure, etc.) and

This is underlined by the IPCC, which concludes that the stabilization of greenhouse gas concentrations at low levels requires 'early investments and substantially more rapid diffusion and commercialisation of advanced low-emissions technologies over the next decades (2000–2030) and higher contributions across abatement options in the long term (2000–2100)'.[6] The new investments required for clean technology development are huge, with estimates ranging from US$ 262 to US$ 670 billion per year.[7]

These findings point to the need to intensify R&D and increase its funding base to ensure the large-scale commercialization of technologies.[8] Whether this requires *new* technologies is debatable. Introducing the concept of 'stabilization wedges', Stephen Pacala and Rob Socolow have argued that a range of existing technologies is already sufficient to reach stabilization of CO_2 concentrations in the atmosphere at around 500 ppm in the first half of the 21st century.[9] Critics have argued that the concept of technology wedges does not properly account for a rapidly expanding demand for energy and, consequently, that policies promoting technological innovation need to be adopted.[10] Furthermore, even though the technologies may already exist, it can be questioned whether the costs for scaling up their use can be justified.[11] In particular, there are concerns that deploying various renewable energy technologies will lead to increased energy prices, which have been kept low by relatively cheap fossil fuels.[12]

There is still uncertainty regarding both the level of mitigation required to avoid dangerous anthropogenic interference with the climate system and the technical and economic viability of specific technologies to achieve emission reductions. According to the *Stern Review*, this uncertainty necessitates the pursuit of a portfolio of clean technologies,[13] rather than a 'silver bullet' technology.[14] The question is, though, what is

---

the 'software' (knowledge required for application of the hardware). See Halsnæs *et al.* 2007, 148; de Coninck 2009, 46–47.

6 IPCC 2007, 68.

7 UNFCCC 2009, para. 94.

8 Hoffert *et al.* 2002, 986.

9 Pacala and Socolow 2004, 968.

10 Victor 2011, 117.

11 Halsnæs *et al.* 2007, 148.

12 Stern 2006, 221.

13 Stern 2006, 358.

14 Halsnæs *et al.* 2007, 148.

the 'right' portfolio? Countries are likely to hold different views depending on their national circumstances.

While the importance of low-carbon and energy efficiency technologies for climate change mitigation is undisputed, their role in the overall climate policy mix is not. For some, scaled-up investments in research, development and deployment should be at the centre of climate policies, as this could possibly kick-start a 'global technology race'.[15] Most scholars, however, argue that promoting technology research, development and deployment is but one part of the climate policy mix and that technology-oriented policies need to be accompanied by other policy instruments, including regulatory and market-based instruments.[16]

This debate is closely related to questions concerning the extent to which governments should be involved in what is 'largely a private-sector endeavour'.[17] While firms play a key role in developing new technologies and subsequently making them commercially available, governments are needed for creating an enabling environment and for providing economic or other incentives to stimulate technological change. A common distinction in this regard is made between 'technology-push' and 'market-pull' policies.[18] The former are aimed at supporting clean technology R&D, whereas the latter establish regulatory or market-based incentives that seek to influence demand for clean technologies.[19] Technology-push measures include, for instance, government subsidies for research on carbon capture and storage, while demand-pull measures include carbon pricing policies or renewable energy technology standards. Again, the general consensus among scholars is that there is no 'either/or' question: both types of policy instruments are needed to address the climate challenge.[20]

An important part of ensuring the large-scale deployment of clean technologies is ensuring that these technologies become widely available. The issue of technology *transfer* thus adds an important international dimension to questions about the development and diffusion of low-emission technologies.[21] Although technology transfer is often framed as

---

15 Galiana and Green 2009, 570; see also Shellenberger *et al.* 2008, 115.

16 Pezzey *et al.* 2008, 107; Neuhoff 2011, 52–54; Victor 2011, 118–119.

17 Halsnæs *et al.* 2007, 155. Governments are not known for being capable of picking technological 'winners'. See Yang and Oppenheimer 2007, 202.

18 Grubb 2004.

19 Halsnæs *et al.* 2007, 156.

20 E.g., Grubb 2004, 130; Jaffe *et al.* 2005, 173; Halsnæs *et al.* 2007, 156; Fischer and Newell 2008, 160.

21 Halsnæs *et al.* 2007, 158.

a North–South issue,[22] the rise of emerging economies such as China and India means that South–South and South–North technology transfer is increasingly taking place.[23] The actual transfer of technologies may occur through different modes: (1) the technology may be incorporated in a product purchased by a recipient country; (2) the recipient country may be licensed to produce such a product; and (3) the recipient country may receive support to produce the product independently. The benefits of the first two modes are clearly with firms in donor countries, whereas a recipient country would likely prefer the third mode.[24]

Technology transfer brings with it several challenges. First, technology transfer requires coordinated efforts between a wide range of stakeholders from the public and the private sector.[25] Donor governments can play an important part, for instance, by providing public funding for technology R&D, and recipient governments could facilitate technology transfer by removing existing market barriers. However, the participation of the private sector is crucial, as firms are ultimately the owners of most clean technologies.[26] In addition, multilateral funding agencies, non-governmental organizations, research and education institutions, and the technology users have an important part to play.[27] Second, who is responsible for technology transfer remains disputed: developing countries generally argue that developed countries, given their ability to pay and historical responsibility for the climate problem, are responsible for facilitating technology transfer, whereas developed countries argue that the developing countries are responsible for creating an enabling environment for investments.[28] Third, there is no agreement on which clean technologies to transfer, as each donor country will likely want to exploit its own competitive advantage. For example, given their relatively large market shares, countries like Germany and Denmark have a vested interest in promoting the transfer of wind energy technologies through the sale of goods (e.g., turbines), whereas a country like France would rather focus on nuclear technologies.[29] Fourth, firms developing technologies usually own the related intellectual property rights, and would

---

[22] E.g., UNFCCC, Art. 4.3.
[23] Brewer 2009, 102–103.
[24] Barton 2007, 3.
[25] IPCC 2000, 4.
[26] Gerstetter *et al.* 2010, 4.
[27] IPCC 2000, 4.
[28] Childs Staley and Freeman 2009, 3.
[29] Gupta 1997, 146. For an overview of selected countries' competitive advantages for several mitigation technologies, see Brewer 2008, 519.

like to see these protected. This has led to concerns that '[intellectual property rights] can in certain circumstances act as a barrier to [clean technology] deployment as they vest in their holders the market power to limit the availability, use, and development of technologies, all of which may combine in higher acquisition cost'.[30] The extent to which this concern holds true in practice is subject to debate, as research shows that the role of intellectual property rights in technology deployment is rather 'peripheral'.[31] Yet for developing countries seeking access to cutting edge technologies, intellectual property rights may still be prohibitive.[32] The precise relationship between intellectual property rights and climate change technologies therefore remains unclear and subject to further analysis.[33]

2.2 The Climate Regime and Clean Technologies

Although the promotion of clean technologies does not play a dominant role in the UNFCCC or the Kyoto Protocol, it is nevertheless the subject of several provisions. The UNFCCC requires its parties to '[p]romote and cooperate in the development, application and diffusion, including transfer, of technologies, practices and processes that control, reduce or prevent anthropogenic emissions of greenhouse gases'.[34] It contains further commitments related to technology transfer, indicating that developed countries need to provide new and additional financial resources, including for technology transfer,[35] and stating that developed countries 'shall take all practicable steps to promote, facilitate and finance, as appropriate, the transfer of, or access to, environmentally sound technologies and know-how to other parties, particularly developing country parties, to enable them to implement the provisions of the Convention'.[36] The latter provision forms a compromise between the position of some developing countries that sought to commit developed countries to transfer technologies on concessional and preferential terms, and the position of developed countries, which rather pursued a commitment

[30] Singh Ghaleigh 2011, 229.
[31] Singh Ghaleigh 2011, 233.
[32] Ockwell *et al.* 2010.
[33] International Centre for Trade and Sustainable Development 2008, 8.
[34] UNFCCC, Art. 4.1(c).
[35] UNFCCC, Art. 4.3.
[36] UNFCCC, Art. 4.5.

towards technology 'cooperation' and sought to safeguard intellectual property rights.[37]

The Kyoto Protocol incorporates the above-mentioned provisions, and expands on them, for instance by also indirectly covering adaptation technologies.[38] Farhana Yamin and Joanna Depledge point to the fact that the Protocol distinguishes more clearly than the UNFCCC between the transfer of technologies that are in the public domain and the creation of an enabling environment for technologies that are privately owned.[39] This distinction reflects an important paradigm shift: rather than viewing technology transfer as an obligation for developed country governments, it draws attention to the domestic situation in recipient (developing) countries and the associated importance of providing incentives for the private sector.[40]

The implementation of the technology provisions of the climate treaties received a boost following a decision adopted in 2001. Following this decision, the UNFCCC discussions have focused on five activities:

(1) *Technology needs and needs assessments*. This includes the identification and analysis of technological priorities for developing countries to provide a basis for subsequent technology transfer.[41]
(2) *Technology information*. This comprises activities aimed at providing information about technical, economic, environmental and regulatory aspects related to technology transfer, as well as information about technology transfer opportunities.[42]
(3) *Enabling environments*. Through this activity, government actions that may enable technology transfer are identified, including fair trade policies, the removal of technical, legal and administrative barriers and enhancing transparency.[43]
(4) *Capacity building*. These activities are largely aimed at building and strengthening capacity in developing countries to allow them to assess and develop environmentally sound technologies.[44]

---

37 Bodansky 1993, 530.
38 Kyoto Protocol, Arts. 10 and 11.
39 Kyoto Protocol, Art. 10(c); see Yamin and Depledge 2004, 306–307.
40 Yamin and Depledge 2004, 307; Tamura 2006, 68.
41 UNFCCC Decision 4/CP.7, Annex, paras. 4–7.
42 UNFCCC Decision 4/CP.7, Annex, paras. 8–11.
43 UNFCCC Decision 4/CP.7, Annex, paras. 12–14.
44 UNFCCC Decision 4/CP.7, Annex, paras. 15–21.

(5) *Mechanisms for technology transfer*. This activity seeks to bring together stakeholders to develop, diffuse and transfer technologies.[45]

To analyse and identify ways to facilitate and advance technology transfer activities, the decision established the Expert Group on Technology Transfer.[46] The Expert Group's broad mandate for promoting technology transfer resulted in a work programme that focused on the five key themes, but also included analyses of information provided in National Communications by parties. In recent years, the Expert Group has also focused on innovative options for financing technology.[47] Although the Expert Group's mandate originally ran only up to 2006,[48] it was extended by one year, and the Group was reconstituted at the 2007 COP in Bali for another five years.[49]

As the UNFCCC's financial mechanism, the Global Environment Facility is tasked with providing financial support for technology-related activities.[50] To help scale up investments, the Global Environment Facility adopted the Poźnan strategic programme on technology transfer in 2008.[51] The strategic programme includes specific funding windows for technology needs assessments, piloting priority technology projects linked to these assessments and dissemination of successful experiences.[52]

The UNFCCC secretariat also plays a role in promoting technology development and transfer. In particular, it has established a technology information system (TT: CLEAR) at the request of the parties. This clearinghouse consists, among other things, of a database of clean technologies and projects, as well as a website collecting information on technology-related activities under the UNFCCC.[53]

In addition to the UNFCCC discussions focusing specifically on technology transfer, the issue also plays an indirect yet important role in the CDM. The Kyoto Protocol does not mention technology transfer explicitly with respect to the CDM, but a follow-up decision mentions

45 UNFCCC Decision 4/CP.7, Annex, paras. 22–23.
46 UNFCCC Decision 4/CP.7, para. 2; Appendix.
47 UNFCCC 2007.
48 UNFCCC Decision 4/CP.7, para. 2.
49 UNFCCC Decision 3/CP.13, para. 3.
50 UNFCCC Decision 4/CP.7, para. 3.
51 UNFCCC Decision 2/CP.14, para. 1.
52 Global Environment Facility 2010, 2.
53 http://unfccc.int/ttclear/jsp/index.jsp (accessed 11 December 2013).

technology transfer as one of the mechanism's goals.[54] There is a burgeoning literature on the contribution of the CDM to technology transfer, fuelled by the idea that it 'is currently the strongest mechanism for technology transfer under the UNFCCC'.[55] The findings of these studies heavily depend on the data sources used (e.g., project design documents drafted by CDM project developers or independently collected data) and the definition of technology transfer (e.g., equipment only and/or associated know-how). While studies using a narrow definition of technology transfer (including the transfer of 'hardware' only) tend to provide cautiously optimistic assessments of the CDM's potential for technology transfer,[56] a more comprehensive definition of technology transfer (including 'software') has led to the conclusion that the CDM's contribution 'can at best be regarded as minimal'.[57] What the various studies do have in common, however, is that they show that the CDM works better for some technologies than for others[58] and that there are differences in the rates of technology transfer among developing countries.[59]

In the absence of clear performance indicators, assessing the performance of the UNFCCC with respect to technology transfer is fraught with difficulties. A report by the UNFCCC secretariat notes that the overall technology transfer framework can generally be deemed effective, though it adds that private sector engagement needs to be improved, and that the level of financial support for the development and transfer of technologies is inadequate.[60]

Technology plays a prominent part in the ongoing climate negotiations. In 2007, 'enhanced action on technology development and transfer' became one of the four building blocks under the Bali Action Plan. The Bali mandate included a consideration of: (1) mechanisms and incentives for scaling up the development and transfer of technology to developing countries; (2) ways to accelerate the deployment, diffusion and transfer of technologies; (3) cooperation on R&D; and (4) the effectiveness of mechanisms for sectoral technology cooperation.[61]

---

54 UNFCCC Decision 17/CP.7, preamble.
55 Schneider *et al.* 2008, 2936.
56 E.g., Haites *et al.* 2006, 346; Seres *et al.* 2009, 4924.
57 Das 2011, 28.
58 E.g., Seres *et al.* 2009, 4924.
59 E.g., Dechezleprêtre *et al.* 2009, 710.
60 UNFCCC 2010b, 47–48.
61 UNFCCC Decision 1/CP.13, para. 1(d)(i–iv).

An initial agreement on a new Technology Mechanism was already part of the 2009 Copenhagen Accord[62] but was fleshed out a year later in the Cancún Agreements.[63] The Cancún decision on technology development and transfer specified that the mechanism is to consist of a Technology Executive Committee and a Climate Technology Centre and Network.[64] The Committee's broad mandate includes the identification of technology needs, addressing barriers to technology development and transfer, cooperation with international technology initiatives and relevant stakeholders and providing guidance and recommendations to enhance the effectiveness of technology development and transfer. The Climate Technology Centre's purpose is to facilitate and coordinate a wider network of (existing) technology initiatives at different levels of governance. The centre and network are intended to provide a flexible framework to support technology development and deployment, especially in developing countries.[65] The mandate for the centre and the network includes the provision of assistance (at the developing countries' request) for identifying technology needs, implementing technologies, and capacity building. In addition, the centre and the network are expected to foster collaboration between public and private institutions on technology development and transfer.[66] With the introduction of the Technology Mechanism, the Expert Group on Technology Transfer ceased to exist.[67]

2.3 Minilateral Clean Technology Agreements and Clean Technologies

A discussion of international cooperation on climate change mitigation technologies that concentrates solely on the UN climate regime is becoming increasingly incomplete. This is primarily due to the rise of international clean technology initiatives involving only a limited number of countries outside of the UNFCCC framework, a development that took off especially in the early 2000s. The prominence of these minilateral initiatives can be partly ascribed to the United States' involvement following its withdrawal from the Kyoto Protocol.[68] Indeed, technology development and cooperation was a key prong of the George W. Bush

[62] UNFCCC Decision 2/CP.15, para. 11; see Gerstetter *et al.* 2010, 10–11.
[63] UNFCCC Decision 1/CP.16, Part IV.B.
[64] UNFCCC Decision 1/CP.16, para. 117.
[65] UNFCCC 2011a; UNFCCC Decision 2/CP.17, Part V.
[66] UNFCCC Decision 1/CP.16, para. 123.
[67] UNFCCC Decision 1/CP.16, para. 124.
[68] McGee and Taplin 2008, 204–207.

Administration's climate strategy.[69] The US is not alone, however.[70] For example, the EU and China concluded a bilateral partnership in 2005 that envisaged intensified cooperation on clean technologies.[71] An analysis of the possible reasons for initiating technology cooperation reveals that the motivations are diverse.[72] Technology initiatives may be pursued because a greater emphasis on clean technology development in climate policy in general is deemed necessary. In addition, *non-UNFCCC* technology agreements may be adopted out of dissatisfaction with the existing UNFCCC/Kyoto approach to promoting (new) technology development and transfer.[73]

It is beyond the scope of this chapter to provide an exhaustive overview of the various international clean technology agreements, but several distinctive characteristics can be identified:

- *Participation.* Different countries participate in the technology initiatives, although most have at least some involvement of industrialized countries.[74] Additionally, some agreements are bilateral whereas others seek to engage a wide range of countries.[75] Furthermore, some initiatives foresee a more prominent role for private sector participants than others. Participation in most international clean technology agreements is limited to a number of countries interested in the technologies covered (i.e., they are generally not multilateral).
- *Technologies covered.* Some agreements focus on one technology. For instance, the Carbon Sequestration Leadership Forum focuses solely on carbon capture and storage.[76] Other initiatives have a broader scope, such as the Renewable Energy and Energy Efficiency Partnership, which covers a variety of renewable energy and energy efficiency technologies.[77]

---

69 Abraham 2004.

70 Tamura 2006, 58; UNFCCC 2010a, 41–51.

71 Chinese Ministry of Foreign Affairs 2006.

72 de Coninck *et al.* 2008, 337–340.

73 Scott Barrett, for instance, argues that the Kyoto Protocol is unable to provide technology 'pull' because of its weak enforcement system, while at the same time also providing no technology 'push' incentives. See Barrett 2003, 393–397.

74 Tamura 2006, 58.

75 UNFCCC 2010a, 41–51.

76 http://www.cslforum.org/ (accessed 11 December 2013).

77 http://www.reeep.org/ (accessed 11 December 2013).

- *Stage of technology cooperation.* Heleen de Coninck and colleagues introduce a typology based on the different stages of technology cooperation, distinguishing between: (1) agreements that promote knowledge sharing, and coordination of research activities and measurement standards (e.g., the Global Methane Initiative); (2) agreements on research, development and deployment activities and their funding (e.g., the European Organization for Nuclear Research); (3) agreements specifying the transfer of technologies (e.g., the Montreal Protocol's Multilateral Fund); and (4) agreements providing mandates, standards and incentives for technology deployment (e.g., the EU Directive on renewable energy sources).[78]
- *Legal nature.* Technology agreements may range from 'statements of "good intentions" to legal contracts'.[79] Some initiatives may be treaty based, especially those containing technology mandates and standards (i.e., type 4 in the classification of de Coninck and colleagues). These initiatives are exemplified by the technology mandates for oil pollution from ships prescribed in the International Convention for the Prevention of Pollution from Ships.[80] Other initiatives, such as the Global Methane Initiative, are explicitly non-legally binding.[81] In between are various shades of grey. For example, Implementing Agreements on various energy technologies concluded under the auspices of the International Energy Agency are not treaty based, but can nevertheless be viewed as a 'contractual relationship' among the Agency's member countries.[82]

The remainder of this chapter focuses on one particular minilateral clean technology agreement, the Asia–Pacific Partnership on Clean Development and Climate. The APP can be characterized as a non-legally binding agreement, led by the governments of a limited number of countries but with private sector participation focusing on a range of technologies, and emphasizing the first stages of technology cooperation (i.e., types 1 and 2 in the classification of de Coninck *et al.*).

---

78 de Coninck *et al.* 2008, 336.
79 Justus and Philibert 2005, 18.
80 de Coninck *et al.* 2008, 347.
81 Global Methane Initiative 2010.
82 International Energy Agency 2003, Art. 1.2.

2.4 The Asia–Pacific Partnership on Clean Development and Climate

Key features of the APP

On 28 July 2005 the APP was officially announced in Vientiane, Laos by the then six participating countries (Australia, China, India, Japan, South Korea and the US).[83] This announcement came four years after US President George W. Bush rejected the Kyoto Protocol; three years after the Australian Prime Minister John Howard did the same; five months after the Kyoto Protocol had entered into force; and three weeks after the G8 Summit at Gleneagles, Scotland, where climate change was high on the agenda.

The announcement of the APP came as a surprise to many observers. The European Commission, for example, was apparently unaware of the Partnership beforehand[84] even though the Gleneagles G8 meeting had just taken place. Most European nations cautiously welcomed the compact. Others, however, were much more sceptical. For example, Friends of the Earth regarded the APP as an attempt to undermine the Kyoto Protocol.[85]

The six founding APP nations comprised some of the world's largest emitters,[86] which was a major rationale given by the founders for creating the APP and for its potential to be effective. In 2005 the countries represented more than half of global CO_2 emissions and worldwide energy consumption, and almost half of global gross domestic product (GDP) and the world population. However, there were also significant differences between the countries: they covered a very broad range of per capita emissions and energy consumption, and the annual income per person varied significantly (Table 6.1). Perhaps one of the most important differences concerned how they viewed their respective obligations to address climate change. All countries in the group, except for the US and Australia, were Kyoto parties when the APP was launched, and had supported the UN climate regime. South Korea had left the G77 in 1996 and had joined the Organisation for Economic Co-operation and Development (OECD). China, and to an even greater extent India, however, vehemently opposed taking on any emission reduction commitments in a future climate agreement, emphasizing the historic responsibility of industrialized countries and their own need to grow and reduce poverty. While members of the APP thus shared certain basic characteristics, such

---

[83] The following draws on Karlsson-Vinkhuyzen and van Asselt 2009.

[84] van der Gaast 2005, 3–4.

[85] Pearce 2005.

[86] B. Fisher *et al.* 2006, 14.

as their geographical location in the Asia–Pacific rim as well as their substantial energy needs and large total emissions, they differed significantly in others, notably their approach to burden sharing in global climate governance.

The APP was created as a non-legally binding 'compact',[87] aiming 'to meet ... increased energy needs and associated challenges, including those related to air pollution, energy security, and greenhouse gas intensities'.[88] Although based on an intergovernmental agreement, it envisaged a key role for the private sector. The Partnership focused on greenhouse gas intensity (i.e., the ratio of greenhouse gas emissions and economic output) and stressed that development and poverty eradication are 'urgent and overriding goals internationally'.[89] Climate change mitigation was thus by implication a subsidiary goal. The participating countries intended to achieve their goals through international cooperation on the development, diffusion, deployment and transfer of clean, efficient, and cost-effective technologies.[90]

The APP in practice

The primary working mode of the Partnership was through sector-specific task forces. To that end, it established task forces for: (1) aluminium; (2) buildings and appliances; (3) cement; (4) cleaner fossil energy; (5) coal mining; (6) power generation and transmission; (7) renewable energy and distributed generation; and (8) steel. Each task force developed an action plan setting out the main objectives and identified activities and projects to be implemented. The main decision-making body of the Partnership was the Policy and Implementation Committee (PIC), whose mandate was to oversee the APP, guide the task forces, and periodically review their work.[91]

In contrast to the UN negotiations, civil society organizations were generally not invited to the international meetings of the Partnership, even though the PIC was in principle open to invite 'relevant governmental, intergovernmental, and non-governmental organizations, where appropriate'.[92] In addition to the private sector, organizations that were invited to attend included the International Energy Agency and the Asian

---

[87] APP Vision Statement 2006.

[88] APP Charter 2006, para 1.1.

[89] APP Charter 2006, para 1.1.

[90] APP Charter 2006, para 2.1.

[91] http://www.asiapacificpartnership.org/english/faq.aspx (accessed 11 December 2013).

[92] APP Charter 2006, para 4.2.

Table 6.1 Selected indicators for the APP nations; all data for 2005[93]

	Share of global CO_2 emissions (%) (excluding LULUCF)	Share of global energy consumption (%)	Share of global GDP (%)	Share of global population (%)
Australia	1.39	1.1	1.1	0.32
Canada	2.03	2.4	2.0	0.50
China	20.26	15.3	9.5	20.19
India	4.44	4.8	4.3	16.94
Japan	4.54	4.7	6.9	1.98
South Korea	1.72	1.9	1.8	0.75
US	21.40	20.9	22.1	4.59
Total	55.78%	51.1%	47.7%	45.27%

	CO_2 emissions per capita (tonnes)	Energy consumption per capita (tonnes oil-equivalent)	GDP per capita (US$)
Australia	18.7	6	31,656
Canada	17.3	8.4	34,972
China	4.3	1.3	4,088
India	1.1	0.5	2,230
Japan	9.8	4.2	30,290
South Korea	9.8	4.4	21,273
US	19.9	7.9	41,813

Development Bank. The participation of non-state actors was slightly more inclusive in the implementation of specific task force projects, but overall the APP was a government-initiated and -driven collaboration with the business sector.

A closer look at the APP task force action plans reveals that most projects were designed to facilitate transboundary interactions between relevant actors in selected sectors in the partner countries. Examples of activities envisaged by the plans include the exchange of specialists, joint

[93] Based on World Resources Institute 2011.

analysis of sectoral mitigation opportunities, the organization of workshops and the development of best practice guides.[94] The project descriptions differed in their level of detail, but some projects also included the demonstration of a new technology, such as the use of clean coal technologies in the cleaner fossil energy task force, and piloting the use of policy instruments in a country, such as labelling energy efficiency of buildings in China. Other activities, for example in the steel and cement sectors, were aimed at the identification of benchmarks and performance indicators. However, none of the action plans set any targets for emission reductions in their sector, and most project activities were unlikely to lead to direct emission reductions. The action plans provided a review of the relevant sectors and areas, and identified barriers to the deployment and diffusion of technologies.

Although public funding was seen as essential for the implementation of the Partnership in terms of leveraging private investments,[95] the pledges for APP activities were limited. The APP's main protagonist, the US government, struggled to secure Congress approval for US$ 42 million (out of the US$ 52 million pledged originally).[96] After a change in government, Australia reduced its pledged funding from A$ 150 to A$100 million (~US$ 157–105 million), spread over five years.[97] When it joined, the Canadian government promised C$ 20 million (~US$ 20 million), spread over three years.[98] Public funding announcements for the other countries were not made, although contributions were at least made indirectly on a small scale through the chairing of task forces and the hosting of APP meetings, workshops and so on. The action plans did not address the funding criticism. For many projects, the plans did not specify how funding would be obtained or whether projects would require additional fundraising. As the funding that was publicly pledged was not sufficient to cover the costs of all the projects proposed, ensuring private sector participation and investments, in combination with support from multilateral investment banks, became necessary to bridge the difference.[99]

After its high-profile announcement in 2005, public attention for the Partnership died away slowly, especially following government changes

---

[94] http://www.asiapacificpartnership.org/english/project_roster.aspx (accessed 11 December 2013).

[95] Pezzey *et al.* 2008, 101–102.

[96] PointCarbon 2006.

[97] Lawrence 2009, 285.

[98] Canada 2008.

[99] Fujiwara 2012, 10.

in the two driving countries, the US and Australia.[100] Although Japan showed increasing enthusiasm for the initiative,[101] it was not enough to save the APP from withering away. In early 2011 the discontinuation of the Partnership was announced and the last meeting of the PIC took place in Bangkok, Thailand, in April 2011.[102] The work of several task forces will continue under the umbrella of other international technology initiatives, including the Global Superior Energy Performance Partnership, and the International Partnership on Energy Efficiency Cooperation.

The nature of the APP

It is not entirely clear whether the APP falls under the consensus regime definition.[103] Jeffrey McGee and Ros Taplin assume so, but also point out that it is uncertain whether the Partnership indeed fulfils the requirements of Stephen Krasner's definition '[g]iven the apparent differences in understanding among [APP] members about key issues such as how the Partnership relates to the UNFCCC and Kyoto'.[104] Put differently, it is unclear whether actors' expectations actually converged in the case of the APP. Still, it can be assumed that this has been the case to at least some extent, for instance with respect to the promotion of clean technology development. For the purposes of this chapter and in line with McGee and Taplin's subsequent analysis, the APP will therefore be considered a regime.

From the perspective of international law, the designation of an international agreement does not determine whether it is classified as a treaty or not.[105] Hence, the usage of the word 'partnership' would not prevent the APP from being covered by international law if the substance of the agreement was legally binding. However, the APP clearly does not fit the definition of a treaty, as the participating countries are explicit in stating that the Partnership should not create any legal rights or duties.[106]

This nevertheless leaves open the question of whether the APP can be considered to be soft law (and whether it has legal effects). Under the binary perspective of law outlined in Chapter 2, there are several arguments that the APP does not qualify as law (and hence also not as

---

100 Karlsson-Vinkhuyzen and van Asselt 2009.

101 van Asselt *et al.* 2009.

102 http://www.asiapacificpartnership.org/english/Default.aspx (accessed 11 December 2013).

103 Krasner 1983, 3. See Chapter 4.

104 McGee and Taplin 2006, 179.

105 VCLT, Art. 2.1.

106 APP Charter 2006, para. 2.1.1.

soft law). The first, formal argument was stated above: the APP cannot be considered a treaty in the sense of the Vienna Convention. The second, related argument is intent based: the founders of the APP clearly did not intend to create any legal obligations or rights. Following a wider definition of soft law, however, there are arguments that the APP can be placed in the grey zone between law and non-law. First, from a formal perspective, the APP's use of a 'Charter', which is usually reserved for constitutive international agreements, gives the APP an almost treaty-like basis. Furthermore, the Charter contains several clauses that are generally found in treaties, such as provisions on amendments, commencement (i.e., entry into 'force'), and termination of the Charter.[107] Second, taking an effect-based approach, the APP sought to influence the behaviour of state and non-state actors, irrespective of its legal characterization. In particular, the Partnership has influenced the behaviour of states vis-à-vis the hard law of the climate treaties. Furthermore, the types of activities which were foreseen in the task forces included explicit norm development such as benchmarking, identifying and promoting best practices, and standards and labelling.

The APP would thus likely fall under the definition of soft law put forward by Kenneth Abbott and Duncan Snidal.[108] While the APP is not a purely political arrangement, it is obviously weakened with regard to its level of legal *obligation*. However, the low level of *precision* also emphasizes its soft character, as it is nearly impossible to verify what conduct complies with the provisions of the Partnership. The pledges made in the APP Charter are of a general, qualitative nature and allow the participating countries to determine for themselves what kind of measures they wish to implement. For example, the objective of the Partnership includes a commitment to 'enhance cooperation to meet … increased energy needs and associated challenges'.[109] Furthermore, the Partnership aims to 'facilitate the development, diffusion, deployment, and transfer of … cleaner, more efficient technologies and practices among the Partners through concrete and substantial cooperation so as to achieve practical results', and to '[p]romote and create enabling environments to assist in such efforts'.[110] What is meant by terms like 'concrete and substantial cooperation', 'practical results', etc. is unclear. Other commitments were slightly more determinate, but were of a highly qualitative nature (e.g., information exchange, engaging the private

---

107 APP Charter 2006, paras. 7 and 8.

108 Abbott and Snidal 2000, 421; see also Abbott *et al.* 2000, 401.

109 APP Charter 2006, para. 1.

110 APP Charter 2006, paras. 2.1.1 and 2.1.2.

sector), which also made it difficult to assess the extent to which the participating nations complied with their commitments. Finally, while the establishment of the PIC and the task forces could be seen as a form of *delegation* of authority, it is much softer than 'high' delegation through courts or quasi-judicial bodies.[111]

The APP would also likely meet the criteria for the definition of 'informal international lawmaking' by Ramses Wessel and colleagues. It concerns cross-border cooperation involving public as well as private actors; is conducted outside of traditional international organizations; involves actors other than the traditional diplomatic actors; and does not result in a formal treaty.[112]

Irrespective of whether the APP falls under these definitions, the main point is that the APP, while not being legally binding and not shaped in the form of an international legal instrument, still sought to influence behaviour, with a potential impact on hard law arrangements, including the climate treaties, as will be discussed in the next section.

3. CONSEQUENCES OF INTERACTIONS

Whether the APP (and minilateral clean technology agreements more generally) and the UN climate regime can be considered competing or complementary depends on the approaches adopted in both venues towards tackling climate change. In other words, how do they define the problem, and what kind of solutions are being promoted?[113]

3.1 Disincentives for Participation in the Climate Treaties

The first interaction relates to the difference in commitments under the APP and the climate treaties, and provides an example of how soft law could serve to undermine hard law. The difference between the APP and the climate treaties not only concerns its legal nature, but also the type of commitments. Whereas the Kyoto Protocol is focused on absolute emission reductions, the APP was framed in terms of greenhouse gas intensity, which meant that absolute emissions could increase even when intensities decrease.[114] Moreover, the commitments were voluntary, and left states with a wide margin of discretion. Although the APP Charter

---

111 Abbott *et al.* 2000, 406.

112 Wessel *et al.* 2011, 8.

113 The following is adapted and updated from van Asselt 2007.

114 McGee and Taplin 2006, 181–184.

called on the member countries to '[a]ssess regularly the progress of the Partnership to ensure its effectiveness',[115] no system to review the progress towards meeting the goals of the Partnership comparable to the Kyoto Protocol's monitoring and reporting system was put in place.

The difference in commitments presented countries in the climate regime with an attractive alternative to participation in the Kyoto Protocol. As a consequence, the APP arguably changed the calculations of actors under the climate regime or made it possible for countries with Kyoto commitments to participate in a climate-related regime without losing credibility.[116] Through the APP, Australia and the United States were able

> to represent and legitimize themselves to their domestic audiences as responsible actors tackling climate change, to vindicate their repudiation of the Kyoto Protocol, to reinforce their case against accepting mandatory emissions reductions in the absence of developing country commitments, to reiterate their commitment to the continued use of fossil fuels, and to justify their voluntarist, technology-led approach, which focuses on reducing emissions intensity rather than aggregate emissions.[117]

Although no country that ratified the Kyoto Protocol fully turned its back on the treaty at the time, it was notable that Canada, a country that faced serious difficulties in meeting its Kyoto commitments, joined the Partnership in 2008.[118] Another country with a Kyoto target, New Zealand, also expressed interest in joining the APP in the early stages.[119] In addition, several other countries, including Mexico, Russia and countries in the Asia–Pacific region expressed their interest in joining the Partnership.[120] This does not mean that the APP would have lured these countries away from Kyoto. Nevertheless, the Partnership, or at least its high-profile status in the first two years of its existence, arguably provided an additional incentive for the Canadian government to express its intention in 2008 not to comply with its Kyoto targets due to excessive economic costs.[121]

---

115 APP Charter 2006, para. 3.1.9.

116 McGee and Taplin 2006, 184.

117 Christoff and Eckersley 2007, 41.

118 Canada 2008. In 2011 Canada formally withdrew from the Kyoto Protocol. See BBC 2011.

119 http://www.mfe.govt.nz/issues/climate/resources/cabinet-papers/mfat-cab-paper.html (accessed 11 December 2013).

120 Peay 2007, 501.

121 Canada 2007, 14–15.

Australia (up to 2007) and the US resisted internationally imposed binding targets in the form of the Kyoto Protocol. The APP presented them with an alternative that met their (economic) interests, as it imposed few economic costs aside from the funding pledged on a voluntary basis.[122] It thus signalled strife in the form of weaker regulation that could trigger an interaction with the climate regime. As Peter Christoff and Robyn Eckersley put it: 'In so far the [APP] lessens the likelihood that Australia and the United States will return to the Kyoto fold and provides a convenient cover against accusations that they are avoiding taking any international responsibility for climate change, it attempts to undermine the efficacy of the UNFCCC.'[123] In short, it can be argued that the APP's presentation of a seemingly credible alternative, combined with its lack of commitments, led to a potential policy conflict with the climate regime, as it provided Kyoto parties like Canada with an additional rationale for not complying with their commitments, and it decreased the chances that non-Kyoto parties like the US would participate in the treaty.

3.2 Inconsistency with the UNFCCC

The second interaction relates to the differences between the APP and the UN climate regime on a more principled level, which sheds doubt on the APP's self-proclaimed consistency with the UNFCCC.[124] While the stated objectives of the APP and the climate regime were overlapping, and the APP Charter provided that '[e]ach Partner will undertake activities … in accordance with … applicable international instruments to which it is a party',[125] its approach diverged significantly from that of the climate regime in three important respects.

First, the APP and the UN climate regime diverged in their approach to differentiation between developing and industrialized countries.[126] The Kyoto Protocol differentiates between Annex B and non-Annex B countries, with only the former having quantified emission limitation and reduction obligations (see Chapter 2). The APP, however, did not differentiate between developing and industrialized nations and at least in its official documents put China and India on an equal footing with the

---

[122] For a comparison of the Kyoto target-and-timetables approach to a technology approach in terms of economic costs, see de Coninck 2009, 215–216.

[123] Christoff and Eckersley 2007, 39.

[124] APP Charter 2006, preamble.

[125] APP Charter 2006, para. 3.2.

[126] McGee and Taplin 2009a, 34–42.

other participating countries.[127] It did not differentiate responsibility for dealing with climate change on the basis of past or current emissions, emissions per capita, or on the basis of capacity, but rather focused on future emissions growth.[128] Under the APP, industrialized countries did not face additional commitments compared to these countries in terms of financial assistance or technology transfer. This lack of differentiation means that '[t]he UNFCCC and Kyoto principle that developed nation responsibility should be differentiated and subject to a higher level of short-term action appears to have been lost'.[129] It should be added that although there was no *formal* differentiation, the *de facto* situation was that many projects were initiated by the industrialized countries of the APP (Australia, Japan, US) and implemented in the developing countries (China, India). Differentiation could thus take place, but only with respect to who would end up paying the expenses for the various projects, leading to a 'rational bargaining' interpretation of the principle of common but differentiated responsibilities.[130]

A second difference between the APP and the climate treaties is that the APP ignored the issue of climate change impacts, and subsequent adaptation to these impacts. Although there is debate over whether and how adaptation and mitigation should be dealt with under a common policy framework,[131] it is conspicuous that there was no discussion at all in the APP about the impacts of climate change or adaptation technologies. By excluding these discussions it becomes impossible to bring in arguments about who is responsible for causing these impacts and, on that basis, who should take the lead in reducing emissions and/or compensate for those impacts. Furthermore, by not including the issues of impacts and adaptation, the Partnership effectively de-linked itself from the overall objective of the UNFCCC to prevent 'dangerous anthropogenic interference with the climate system'.[132]

Third, there are differences in the way the APP and the UN climate regime approached the participation of non-governmental stakeholders. The UNFCCC explicitly provides for the admission of non-governmental

---

127 The Republic of Korea is also a non-Annex B country. However, its GDP per capita is higher than some countries with commitments. Therefore, the differentiation arguments made here mainly apply to China and India.

128 Christoff and Eckersley 2007, 38.

129 McGee and Taplin 2006, 188.

130 McGee and Taplin 2009a, 41.

131 Berkhout 2005, 383–385.

132 UNFCCC, Art. 2.

organizations as observers to sessions of the treaty bodies.[133] Non-governmental organizations have actively participated in the climate negotiations throughout the development and implementation of the regime.[134] Like the UNFCCC, the Partnership aimed to integrate 'development banks, research institutions, and other relevant governmental, intergovernmental, and non-governmental organisations, as appropriate'.[135] However, complaints from many environmental non-governmental organizations about the transparency of the process point to the fact that the organizations mentioned in the Charter were only admitted on an ad hoc basis, without the use of any objective criteria, and that there was no systematic procedure for their admission. Admitting other observers, including non-governmental organizations and research organizations, could have led to an enhancement of the legitimacy and environmental integrity of the APP, as these actors could help keep track of the implementation of the Partnership.[136]

The APP countries, and in particular the US, were well aware of the potential influence they could exert on the climate regime. At least formally they made it seem that there would be no incompatibility between the APP and the climate treaties. However, merely claiming 'complementarity' or 'consistency' is insufficient. The Kyoto Protocol and the Partnership cannot be considered complementary '[i]f one policy obstructs or undermines the effectiveness of another'.[137] This may be partially the case, as it was possible for the Partnership to obscure some aspects that many consider integral to the UN climate regime (differentiation, impacts and adaptation, and stakeholder participation). The state practice of countries participating in the APP that were also a party to the UNFCCC and/or the Kyoto Protocol could hence run counter to the provisions of the treaties they had signed and ratified, pointing to a behavioural interaction that led to a potential policy conflict between the APP and the UN climate regime.

3.3 Implementation Support for the Climate Treaties

The third interaction underlines the synergies between non-UNFCCC minilateral clean technology agreements and the UN climate regime. Even though there was no legal obligation to do so, the APP countries

---

133 UNFCCC, Art. 7.6.

134 Gulbrandsen and Andresen 2004.

135 APP Charter 2006, paras. 3.1.7 and 4.2.

136 Lawrence 2007, 201.

137 McGee and Taplin 2006, 178.

started a wide variety of projects that could support the implementation of the UNFCCC, the Kyoto Protocol and any follow-up agreement.

First, various task force action plans envisaged data collection on, for example, greenhouse gas emissions, emissions intensity, projections and reductions potentials, which could enhance the quality of reporting under the climate treaties for both Annex I and non-Annex I countries. Both groups need to submit National Communications that include greenhouse gas emission inventories.[138] Emissions data collection could thus provide support in drafting the various reports under both the UNFCCC and Kyoto Protocol.

Second, many projects concerned the exchange of information between governments and business actors in the APP countries. These activities could help build capacity to implement climate-friendly policies and projects in countries such as China and India, and could facilitate the transfer of environmentally sound technologies. In this way, the APP could provide a way to support 'endogenous capacities' for technology development, as required under the UNFCCC.[139] At the project level, the APP envisaged capacity building through, for example, training plant engineers in China and India in how to improve their energy efficiency. At the government level, countries could learn from each other's successful policies.

Third, the APP helped the developed countries implement their commitments under the climate treaties, especially those related to technology development and transfer. For Japan, for instance, the Partnership presented a no-regrets strategy, possibly contributing to emission reductions at home in pursuance of its domestic emission reduction targets.[140] As the country is known for having one of the highest energy-efficiency rates in the world, it also provided a potential way to diffuse existing energy-efficient technologies into the other countries, and could provide a contribution to the achievement of (future) targets of the other countries (see Section 6 below).

This type of interaction corresponds to the view that soft law and hard law can be complementary and that there are synergies between the climate regime and minilateral clean technology agreements. Although the UN climate regime has put in place a broad framework to promote technology development and transfer, this framework needs to be realized

138 UNFCCC, Art. 12.1.

139 UNFCCC, Art. 4.5; see also Kyoto Protocol, Art. 10(c).

140 van Asselt *et al.* 2009, 332–333.

in practice and requires the participation of a broad group of stakeholders. This example illustrates how the APP and its project-oriented approach could contribute to that.

3.4 Testing Ground for the Future Architecture of Global Climate Governance

A fourth way through which the APP and the UN climate regime could interact is if the Partnership fulfilled the role of 'policy model'[141] from which decision makers in the climate regime could learn. This cognitive interaction would mean that policymakers in the climate regime could include elements of the APP in a future framework. Several features of the Partnership are important in this regard: its minilateral approach; the key role for business actors; non-legally binding (voluntary) commitments; its sectoral approach; the focus on greenhouse gas intensities rather than absolute emission reductions; enhanced incentives for technology transfer; and strengthened links with other policy objectives (energy security, local air pollution, etc.).

In UNFCCC meetings in 2007–2008, the industrialized countries participating in the APP made various references to the Partnership, in particular advocating the 'sectoral approach' it adopted as well as its voluntary, bottom-up approach.[142] Japan was particularly active in citing the APP as a key example of its proposed sectoral approach, even though the sectoral approach it suggested in the UN climate negotiations did not match the model of the APP.[143] Sectoral approaches found their way into the negotiations, notably in the Bali Action Plan.[144] Furthermore, other elements of the APP, such as its voluntary, bottom-up approach, its lack of differentiation between developed and developing countries, and its minilateral process, are also increasingly prevalent in the UN climate negotiations, and are reflected in the Copenhagen Accord and the Cancún Agreements, even if they are still strongly contested (see Chapter 2).[145] This is not to say that the APP is the *reason* that these elements started to play a more important role in the climate negotiations. However, the Partnership provided proponents of an alternative global climate governance architecture with an additional venue in which they could market its

141 Gehring and Oberthür 2006a, 328–330.
142 Vihma 2009, 255.
143 van Asselt *et al.* 2009, 326.
144 UNFCCC Decision 1/CP.13, para. 1(b)(iv).
145 Rajamani 2011, 519.

purported benefits. McGee and Taplin argue that this 'discursive coalition' used the APP and other forums, such as the Major Economies Process and the Asia–Pacific Economic Cooperation summit to contest future commitments under the climate regime.[146]

The APP can thus be seen as a testing ground for a future climate governance architecture. Whether this leads to conflicts or synergies is unclear though. The cognitive interaction between the APP and the UN climate regime can be seen as a case of soft law being fertile testing ground for future hard law, with the APP strengthening the climate regime. Yet it may equally show how soft law may contribute to the softening of existing hard law, meaning that the APP weakens the climate regime. It remains to be seen whether the approach advocated by the APP, and mirrored by the Copenhagen Accord and the Cancún Agreements 'may in fact engage a growing number of States in individual mitigation efforts that will lay the groundwork for a future formal treaty as a follow-up to Kyoto'.[147] The jury thus remains out on the consequences of this potential interaction.

4. INTERACTION MANAGEMENT: LEGAL TECHNIQUES

The opportunities for applying legal techniques to the interactions between the APP and the climate regime are particularly limited. Generally, it can be said that non-legally binding agreements are not governed by international law.[148] This means, as Dan Bodansky remarks, that 'many of the issues addressed by the law of treaties do not have a counterpart for non-legal agreements'.[149] Indeed, the Vienna Convention on the Law of Treaties, or conflict resolution principles such as *lex specialis* are not applicable to resolving the 'antagonistic' interaction between the soft law of the APP and the hard law of the climate treaties, since there will not be any normative conflict in the strict legal sense. Treaty interpretation rules are in principle also not applicable, as the APP cannot be regarded as a treaty, although these rules may be nevertheless be used to interpret provisions of the APP Charter. For example, McGee and Taplin use the VCLT's interpretation rules to interpret the use of the

146 McGee and Taplin 2009b, 232–234.
147 Pollack and Shaffer 2012, 262.
148 Schachter 1977, 301.
149 Bodansky 2009b, 156.

term 'consistent with the UNFCCC' in the APP Charter.[150] Particularly from the perspective of a government that participates in technology agreements as well as the climate treaties, treaty interpretation may thus be useful for enhancing understanding of the Partnership, and could also allow for a more harmonious view of the different regimes.

5. INTERACTION MANAGEMENT: INSTITUTIONAL COORDINATION

5.1 Existing Institutional Coordination

Institutional coordination between the UN climate regime and the APP was practically non-existent, with links between the two primarily being highlighted through submissions by the countries in the climate regime that also participated in the APP.[151]

Nonetheless, there is potential for parties to the climate treaties to link non-UNFCCC initiatives to the climate regime in a future climate agreement. A first sign in this direction can be found in the Cancún Agreements. The decision encourages parties to engage in bilateral and multilateral cooperative activities on technology development and transfer.[152] In other words, parties to the UNFCCC acknowledge and encourage a diversity of initiatives. Furthermore, the decision mandates the Technology Executive Committee to establish cooperation with relevant international technology initiatives, stakeholders and organizations and to promote coherence and cooperation across various technology activities within and outside of the UNFCCC.[153]

There is already some experience of institutional coordination with non-UNFCCC clean technology initiatives. The Climate Technology Initiative is a good example. This initiative of 11 OECD countries[154] was established at COP1 in 1995, following exploratory work by the International Energy Agency, and has been the subject of one of the Agency's Implementing Agreements since 2003. Participants in the initiative have worked in close collaboration with both the UNFCCC secretariat and the Expert Group on Technology Transfer, and the work

---

150 McGee and Taplin 2009a, 21.
151 Vihma 2009.
152 UNFCCC Decision 1/CP.16, para. 116.
153 UNFCCC Decision 1/CP.16, para. 121(f).
154 http://www.iea.org/techno/iaresults.asp?id_ia=7 (accessed 11 December 2013).

of the initiative has been linked to the five technology themes identified by the Marrakech Accords.[155] One of the examples of cooperation with the Expert Group is the Private Financing Advisory Network, which seeks to identify projects in developing countries for which it might be possible to find private sector finance, and facilitate access to such finance.[156] Another example is a handbook for technology needs assessments, which was jointly produced between the Climate Technology Initiative, the Expert Group on Technology Transfer, and the UNFCCC secretariat.[157] The example of the Climate Technology Initiative thus shows that informal and ad hoc cooperation and coordination is possible between international clean technology agreements and bodies of the UN climate regime.

5.2 Enhancing Institutional Coordination

Formalizing institutional coordination

A next question is whether institutional coordination between the UN climate regime and minilateral clean technology agreements could or should be formalized, with a view to encouraging more structural coordination. Although the Cancún mandate does not mention the possibility, parties to the UNFCCC could conclude a Memorandum of Understanding with organizations facilitating clean technology cooperation. Such a memorandum could request the administrative bodies of various minilateral clean technology agreements to report on their activities and mandate ongoing cooperation with UNFCCC bodies such as the secretariat or the new Technology Executive Committee. Memoranda of Understanding have been used before to initiate cooperation under the UNFCCC, for instance with the Global Environment Facility.[158] However, a difference between the Global Environment Facility and most minilateral initiatives is that as the financial mechanism of the treaty, the Facility already had a formal link to the UNFCCC.[159]

There are several arguments for at least a similar level of formal cooperation and coordination between the UNFCCC and minilateral initiatives.[160] First, it could ensure a higher degree of transparency

---

155 E.g., UNFCCC Decision 4/CP.7, Annex, para. 10(a).
156 http://www.cti-pfan.org/ (accessed 11 December 2013).
157 United Nations Development Programme 2010.
158 UNFCCC Decision 12/CP.2, Annex.
159 UNFCCC, Art. 11.
160 Carlino *et al.* 2007, 26.

regarding activities taking place under such initiatives, providing a clear signal that they are intended to complement the efforts under the climate regime.

A second argument in favour of coordination is to keep track of how the different minilateral technology agreements contribute to overall mitigation objectives. For instance, coordination could facilitate an assessment of the extent to which the emission reductions achieved by different technologies could lead to a stabilization of greenhouse gas concentrations.[161] Furthermore, by keeping track of activities the double counting of mitigation actions under different agreements could be avoided.[162]

Third, coordination could avoid duplication of work: experiences with the development and transfer of similar technologies could be shared among the initiatives, thereby ensuring more efficient international technology cooperation. The technology and project databases maintained by the UNFCCC secretariat already partially fulfil this function, but the databases could still be enriched by the experiences from minilateral clean technology agreements not reported by parties in their National Communications. Coordination could further prevent duplication of work by promoting the use of common accounting standards and monitoring and reporting formats.

Fourth, coordination could ensure that technology development and transfer better correspond to technology needs. In this regard, a review by the Chair of the Expert Group on Technology Transfer identifies a number of deficits, including a limited focus on least developed countries, a strong emphasis on the energy sector (rather than industry, transport, forestry and agriculture, etc.) and a dominance of mitigation (rather than adaptation) technologies.[163] Moreover, to ensure that the development and transfer of technologies is effective on the ground, they need to be linked to the assessment of the technology needs of recipient countries. Such technology needs assessments as well as the identification of gaps in technology cooperation would likely transcend the more restricted scope of most minilateral clean technology agreements, especially those focused on a limited number of technologies.

Fifth, coordination could help to channel public funding to the various initiatives, most notably through the Green Climate Fund established by

---

161 Cf. Pacala and Socolow 2004.

162 Murphy *et al.* 2005, 14; de Coninck 2009, 218.

163 UNFCCC 2010a.

the Cancún Agreements.[164] The fund may (over time) include a specific thematic funding window for technology development and transfer.[165] At the moment, the Clean Technology Fund, one of the two multi-donor Climate Investment Funds established by the World Bank, fulfils this function on an interim basis. However, according to the Bank, this fund 'will take necessary steps to conclude its operations once a new financial architecture [under the UNFCCC] is effective'.[166] The funding pledged for the Clean Technology Fund as of April 2013 was US$ 4.936 billion,[167] and should be expected to grow with the establishment of the Green Climate Fund. The case of the APP shows that even though a multitude of projects may be developed, the necessary public funding is not always forthcoming.[168] Coordination of the concrete activities developed under minilateral clean technology agreements and the work of the UNFCCC could overcome this obstacle.

The strengths and functions of the UN climate regime and minilateral clean technology agreements

Enhanced institutional coordination between the UN climate regime and minilateral initiatives is likely to be most effective if it aims at making use of their respective strengths. While I do not intend to provide an exhaustive overview of the strengths and weaknesses of the UNFCCC and minilateral clean technology agreements, some key strengths of each are presented in Table 6.2.

The general strengths of the minilateral clean technology agreements outside the UNFCCC include notably the high participation rates of those actors most relevant for, and interested in, specific mitigation technologies.[169] This not only refers to states interested in developing specific technologies, but the hybrid mode of governance often used also allows for participation of the private sector (i.e., the actors owning, developing and buying technologies). These actors have an interest in participating in clean technology agreements in order to reap the benefits from moving

---

[164] Not all technologies will require public funding, and public support is likely to be most useful if it is used to lower the risk of (private) investment in certain technologies. See Benioff *et al.* 2010, 33–35.

[165] Decision 3/CP.17, Annex, paras. 37–39.

[166] World Bank 2008b, paras. 56–57.

[167] http://www.climatefundsupdate.org/listing/clean-technology-fund (accessed 1 July 2013).

[168] Pezzey *et al.* 2008, 105–106.

[169] de Coninck *et al.* 2008, 336, 350.

Table 6.2 Strengths of the UNFCCC and minilateral clean technology agreements[170]

UNFCCC	Minilateral clean technology agreements
• Links to global climate change mitigation objectives • Relatively high transparency • Links to funding mechanisms • Established institutional framework • Technology information system to share experiences	• High participation rates (of state and non-state actors interested in a specific technology) • Involvement of private sector (technology owners) • Facilitates quick decision making • Flexibility to adapt to technological developments

first in developing technologies. Furthermore, the discussion of clean technology development in a non-adversarial, cooperative setting generally allows for speedier decision making.[171] Finally, the use of informal instruments increases the flexibility to adapt to new technological developments, avoiding the need to resort to potentially lengthy amendment procedures.

The next question is: which functions related to technology development and transfer could be best fulfilled by either the UNFCCC or the minilateral clean technology agreements?

Perhaps the least controversial role that the UN climate regime could play is intensifying the functions that it has carried out already. This includes, first, its information- and knowledge-sharing function, which it has in part fulfilled by acting as a clearing house for technology cooperation, and continuing the development and compilation of technology needs assessments. Second, the UN climate regime could likely play a role by supporting technology hubs and innovation networks.[172] The Climate Technology Centre and Network established by the Cancún Agreements have clear potential to link national and regional centres of technology expertise to each other. Third, the UNFCCC can continue to play an important role in terms of capacity building in developing countries, particularly through its emphasis on technology needs assessments. Fourth, as indicated above, the climate regime could form the

---

[170] Based on Carlino *et al.* 2007, 21; Tawney and Weischer 2011.

[171] Keohane and Victor 2011, 18; Victor 2011, 242–243.

[172] Tawney and Weischer 2011, 4.

focal point for the provision of public finance for some technologies through the Green Climate Fund. A more far-reaching change would be to also make the UNFCCC a forum to establish criteria for financing initiatives through the Green Climate Fund, and through the operationalization of provisions on monitoring, reporting and verification. Such a decision would undoubtedly be politically charged, as it would give the UNFCCC, with its one-country-one-vote structure, authority over considerable financial resources provided by developed countries. However, it could also lead to an accurate and ongoing assessment of the contribution of the various initiatives to the Convention's overall objectives.

What the UN climate regime is unlikely to be able to do is to influence the choices about the kinds of clean technologies to develop, given that political preferences based on national circumstances will likely dictate such choices. Nevertheless, it may be possible that a UNFCCC-based body could at least prevent 'powerful states [from] exploiting their position and putting their economic interest in a specific technology above the general interest of environmental effectiveness'.[173] This relates to the strength of the UNFCCC to link specific actions to an overall (mitigation) objective. Whether states are willing to be constrained in their technology policy choices remains to be seen.

6. AUTONOMOUS INTERACTION MANAGEMENT

In the event that enhanced institutional coordination between the UN climate regime and minilateral clean technology agreements proves to be challenging, it is still possible that countries participating in both live up to their commitments under the climate treaties. The potential outlined above for minilateral clean technology agreements to complement the UNFCCC's work on technology development and transfer shows that countries can participate in both forums in pursuit of the same objective. Furthermore, the extent to which minilateral initiatives are in conflict with the UN climate regime could be mitigated by the way that individual countries profile these initiatives, as it was precisely the way the APP was profiled in 2005 by its main protagonists that gave rise to the perception that it was competing with the climate treaties.

An interesting case of autonomous interaction management is how Japan sought to navigate between the APP and the UN climate regime

[173] de Coninck 2009, 218.

during the APP's existence.[174] At the time of the formation of the Partnership, Japan was placed in a unique position, as it was the only country that also had an emission reduction target under the Kyoto Protocol. Australia and the US were also among the developed countries in the UNFCCC, but had not (yet) ratified the Kyoto Protocol. China, India and South Korea had ratified the Protocol, but did not have legally binding emission reduction targets. Japan, in contrast, had committed itself to reduce its emissions by 6 per cent from 1990 levels between 2008 and 2012. Japan's position was also noteworthy for other reasons. First, it had a special interest in the success of the Kyoto Protocol, with the treaty being named after its former capital city.[175] Second, Japan played a critical role in forging consensus on the Bonn Agreement in 2000–2001, when there was mounting concern regarding the future of the Kyoto Protocol due to the United States' decision not to ratify.[176] Third, the country has historically attempted to maintain good relations with the main non-party to the Kyoto Protocol, the US.[177] Lastly, Japan has interests in exerting political leadership in Asia.[178] So while Japan had decided to join the APP, it was clear that it could not easily turn its back on the Kyoto Protocol. Instead it sought to highlight how the Partnership could be used to support the UNFCCC and to provide a model for future climate governance.

The APP displayed a number of characteristics that fit well with the Japanese position in the post-2012 discussions in the UN climate regime: (1) its focus on emissions in specific heavy-emitting sectors that are subject to international competition; (2) the emphasis on technological innovation and the transfer of clean technologies and practices; and (3) the participation of major emitters (China, India, US) in mitigation efforts. Nevertheless, the APP also lacked some of the elements that Japan embraced in the UN context. Notably, APP activities were not linked to any long-term objective for global emission reductions. Furthermore, the sectoral approach of the APP was different from that advocated by Japan in the UNFCCC discussions. In the APP, sectoral activities were of a voluntary nature, whereas Japan's proposals in the climate regime sought to use sectoral analysis as the basis for a country's (binding) commitments.[179]

---

174 The following draws on van Asselt *et al.* 2009.
175 Tiberghien and Schreurs 2007, 88.
176 Broadbent 2002.
177 Matsumura 2000, 5.
178 Matsumura 2000, 66.
179 UNFCCC 2008, 33–34.

For Japan, the APP thus did not present the same kind of alternative to the Kyoto Protocol that the US and Australia had in mind. Still, the country became one of the APP's most enthusiastic supporters at the end of the Partnership, after it had diminished in importance following changes in government in both the US and Australia. Japan's position can be explained not only by the inter-ministerial conflicts that have characterized its domestic politics, but also by the strong private sector support for the APP, which included industries directly involved in some of the APP task forces.[180] Its ambivalent position towards the APP meant that Japan did not implement the Partnership in a way that would threaten compliance with its commitments under the climate treaties. At the same time, however, it allowed the country to strategically use the Partnership as a model for international climate governance. This shows how autonomous interaction management through implementation by states that are aware of the regime interactions can address potential conflicts and enhance synergies.

7. CONCLUSIONS

Minilateral clean technology agreements such as the APP pose a challenge to both political scientists and international lawyers as they fall outside traditional definitions and concepts. This novelty should not be used as a reason to disregard the emergence of such institutional arrangements, as these new forms of climate governance may affect the climate regime and have implications for the substance and form of a future climate agreement.

In terms of consequences of regime interactions, the chapter has highlighted two potential policy conflicts. First, because the commitments of the APP deviated from those agreed upon in the Kyoto Protocol in terms of legal nature and precision, incentives were created for countries to steer away from a treaty that contains such binding commitments, highlighting a behavioural interaction resulting in a policy conflict. Second, the APP and the climate regime diverged on several key issues: differentiation between industrialized and developing countries; impacts and adaptation; and the participation of a wide range of stakeholders. These differences point to another behavioural interaction that led to a potential policy conflict with respect to the principles underlying the climate regime and the APP.

---

[180] van Asselt *et al.* 2009.

Nonetheless, the implications of minilateral clean technology agreements are not necessarily negative, and the chapter has drawn attention to behavioural interactions that could produce actual and potential synergies between the APP and the climate regime to illustrate this point. In particular, the APP's activities in the area of data collection, knowledge exchange, and capacity building have supported the climate regime's processes for promoting technology cooperation, and similar activities under other minilateral initiatives may continue to do so in the future. In addition, in a case of cognitive interaction, the APP formed one of the testing grounds for a future global climate governance architecture, although it remains to be seen to what extent its features will be integrated in the UN climate regime, and whether that would affect the effectiveness of the latter.

Moving on to interaction management, the chapter has first pointed out that the scope for legal techniques in this case is practically non-existent, as one of the regimes is not based on a treaty governed by international law. It has further shown how institutional coordination between the UN climate regime and minilateral clean technology agreements is slowly taking shape. Following an examination of the rationales for enhanced institutional coordination, it has argued that enhanced institutional coordination would likely be most successful if it made use of the respective strengths of the UNFCCC and non-UNFCCC initiatives in the area of technology development and transfer.

While the importance of the APP has decreased since its termination in 2011, it is by no means evident that future climate governance will be centred upon the UN process. Any multilateral climate agreement will most likely be accompanied by flanking minilateral institutions involving countries and private actors that are not satisfied with the negotiation outcomes, either because they are perceived to be not ambitious enough or because they are seen as overly ambitious.[181] Moreover, minilateral initiatives outside of the UNFCCC will remain relevant as they could influence the shape of UN climate governance at the cognitive level. Already, the sectoral approach advocated in the APP and other initiatives has reached the negotiation agenda, even though it is not yet fully understood what is meant by it. It is certainly possible that other elements of the APP will figure prominently in the ongoing climate negotiations.

[181] A case in point is the Climate and Clean Air Coalition, which emerged in 2012 as an initiative involving states and non-state actors with a view to tackling short-lived climate pollutants, an area in which the UNFCCC has only made limited progress: http://www.unep.org/ccac/ (accessed 11 December 2013).

In this regard, coordination of the activities under minilateral initiatives and under the UNFCCC process could ensure that they work towards the common objective of avoiding dangerous climate change. Coordination would arguably lead to a more systematic evaluation of the various barriers to technology transfer and offers a way to deal with them not only at the regional and minilateral, but also at the global and multilateral, level. Furthermore, a system that monitors the contribution of minilateral initiatives to the common objective would ensure a higher degree of transparency of their activities and make clear how they would contribute to compliance with commitments under the climate regime. More importantly, it would provide a clear signal that these minilateral approaches are not meant as competitors to the multilateral climate regime, but rather as complements.

7. The UN climate regime and the Convention on Biological Diversity

1. INTRODUCTION

This second case study zooms in on the interactions between the UN climate regime and the Convention on Biological Diversity as a case of interactions between two treaty-based regimes within the field of international environmental law and governance. Focusing specifically on the role of forests at the nexus of climate and biodiversity governance, the chapter seeks to analyse the consequences of the interactions between the climate and biodiversity regimes. It further aims to show how legal techniques and different ways of institutional coordination could ensure that international policies promoting climate change objectives and policies reducing biodiversity loss could be mutually supportive.[1]

The relationship between the climate and biodiversity treaties goes back to their very origins. The UNFCCC and CBD were negotiated in parallel during the heyday of modern international environmental law-making. Both treaties were submitted for adoption to the Earth Summit in Rio de Janeiro in 1992, and were heralded as an important achievement of this landmark conference. While the 'Rio Conventions', together with Agenda 21, were the most notable positive outcomes of the Earth Summit, countries at the meeting failed to reach agreement on a global forest convention.[2] With a legally binding agreement on forests out of reach, countries settled for the adoption of the non-legally binding 'Forest Principles'[3] and the inclusion of a chapter on deforestation in Agenda 21.[4] Since 1992, various initiatives have emerged in the area of global forest governance, which have kept the hope of a global forest

[1] This chapter is an updated and shortened version of van Asselt 2012b.
[2] D. Humphreys 1996, 83–104.
[3] UNGA 1992b.
[4] UNGA 1992a, Chapter 11.

convention alive.[5] In addition, as this chapter will highlight, actors in existing international environmental regimes, including the climate and biodiversity regimes, have become increasingly aware of the links between forests and their respective subject matters.[6]

2. THE GOVERNANCE OF CLIMATE CHANGE AND BIODIVERSITY

2.1 Climate Change and Biodiversity Loss

The linkages between two of the main environmental threats facing the world today, climate change and biodiversity loss, are manifold and complex.[7] In the first place, climate change is a major threat to the conservation of biodiversity and already has negative impacts on ecosystems, species, genetic diversity and ecological interactions.[8] This includes impacts on the distribution of ecosystems, for instance by inducing a poleward shift of ecosystems,[9] and changing the composition of ecosystems, for instance through the introduction of invasive alien species.[10] Another important connection is that ecosystems with high biological diversity are generally more resilient to climate change and variability than impoverished ecosystems. Hence, if other pressures on biodiversity decrease,[11] it is more likely that ecosystems will adapt naturally to climate change.[12] Furthermore, biodiversity can support humans in their efforts to adapt to climate impacts.[13] For instance, coastal ecosystems can strengthen coastal defence systems to prevent floods and

---

[5] D. Humphreys 2006, 114–115. It is beyond the scope of this chapter to discuss the multitude of public, private and public–private initiatives in global forest governance. See McDermott *et al.* 2007; Rayner *et al.* 2010.

[6] D. Humphreys 2006, 190.

[7] IPCC 2002; Secretariat of the CBD 2003; Secretariat of the CBD 2009a; Secretariat of the CBD 2009b.

[8] Secretariat of the CBD 2009a, 12–29.

[9] Parmesan and Yohe 2003.

[10] Rahel and Olden 2008.

[11] Other notable threats to biodiversity include habitat loss, overexploitation, pollution, and the invasion of alien species. See Millennium Ecosystem Assessment 2005, 96–99.

[12] Secretariat of the CBD 2003, 78.

[13] Secretariat of the CBD 2009a, 73–74.

erosion.[14] Finally, ecosystems play an important role in the carbon cycle by either taking up (sequestering) or emitting carbon.[15]

The complex relationship between climate change and biodiversity is especially pertinent in the case of forest ecosystems. Forests are an important component of the world's biological diversity and play a key role in maintaining global biodiversity. Intact forests will likely be more resilient to climate impacts, making protection of such forests a sound adaptation strategy.[16] At the same time, forests are important from the perspective of climate change mitigation as they form either net carbon sinks or sources of emissions. Young growing trees act as sinks by absorbing carbon dioxide from the atmosphere. Most CO_2, however, is stored in old-growth forests, which form vast reservoirs of carbon over a long period. When forests or harvested wood products are burned or decompose, the biomass loses its function as a sink and becomes a source of carbon.[17] Varying estimates indicate that tropical deforestation and forest degradation accounts for about 12–20 per cent of global CO_2 emissions.[18] Measures to avoid deforestation, as well as afforestation, reforestation and forest management hence have a significant potential for climate change mitigation, but their impacts on biodiversity vary.[19]

2.2 The Climate Regime and Biodiversity

General provisions

The objectives of the climate and biodiversity regimes can generally be said to be converging, as both are concerned with the conservation of ecosystems. Although the climate and biodiversity treaties do not explicitly refer to each other, the texts include indirect connections. The UNFCCC's objective of stabilizing greenhouse gases at non-dangerous levels is to be achieved 'within a time-frame sufficient to allow ecosystems to adapt naturally to climate change'.[20] Parties to the UNFCCC are also committed to 'promote and cooperate in the conservation and enhancement … of sinks and reservoirs … including biomass, forests and oceans as well as other terrestrial, coastal and marine ecosystems'.[21]

---

14 Secretariat of the CBD 2009a, 54–56.
15 Secretariat of the CBD 2009a, 93–97.
16 Secretariat of the CBD 2009a, 62–63.
17 Secretariat of the CBD 2009a, 48.
18 IPCC 2007, 36; van der Werf *et al.* 2009, 737.
19 Secretariat of the CBD 2009a, 102–104.
20 UNFCCC, Art. 2.
21 UNFCCC, Art. 4.1(d); see also Art. 1.1 and Art. 4.8.

Furthermore, in adopting climate change mitigation measures, parties need to 'employ appropriate methods, for example impact assessments … with a view to minimizing adverse effects … on the quality of the environment'.[22]

Unlike the UNFCCC, the Kyoto Protocol does not refer explicitly to biodiversity or ecosystems, but it calls on its parties to implement policies and measures, including the protection and enhancement of sinks, 'taking into account [their] commitments under relevant international environmental agreements'.[23] The same provision also calls on parties to implement measures for the 'promotion of sustainable forest management practices, afforestation and reforestation',[24] although parties so far have not defined what is meant by 'sustainable' in this context.[25] The Protocol also demands that parties implement policies and measures in such a way as to minimize the effects, including environmental effects, on other parties.[26] Finally, the COP is instructed to assess the environmental impacts of measures taken pursuant to the Protocol.[27] While the environmental effects could refer narrowly to the effect on greenhouse gas emissions only, the more appropriate interpretation would broadly include all kinds of environmental impacts, including those on biodiversity and ecosystems.

The climate regime and forests

Forests have played a prominent role in the climate regime in the context of discussions about removals and emissions from land use, land-use change, and forestry (LULUCF). Forests are considered to be first and foremost a means of climate change mitigation through their function as carbon sinks (and, conversely, their potential as source of emissions). The main commitment contained in the UNFCCC in this regard is that all parties must report annual emissions by sources and removals by sinks.[28] Furthermore, all parties must promote the sustainable management of all sinks and reservoirs, including forests.[29]

The Kyoto Protocol put the use of carbon sinks high on the agenda by opening up the possibility of using sinks to meet the emission reduction

---

22 UNFCCC, Art. 4.1(f).
23 Kyoto Protocol, Art. 2.1(a)(ii).
24 Kyoto Protocol, Art. 2.1(a)(ii).
25 Pontecorvo 1999, 731; Sagemüller 2006, 208.
26 Kyoto Protocol, Art. 2.3.
27 Kyoto Protocol, Art. 13.4(a).
28 UNFCCC, Art. 4.1(a).
29 UNFCCC, Art. 4.1(d).

targets agreed upon by industrialized countries. Article 3.3 of the Protocol states that

> net changes in greenhouse gas emissions by sources and removals by sinks resulting from direct human-induced land-use change and forestry activities, limited to afforestation, reforestation and deforestation since 1990, measured as verifiable changes in carbon stocks in each commitment period, shall be used to meet the commitments [of industrialized countries].[30]

In addition, Article 3.4 determines that rules concerning the use of LULUCF activities other than afforestation, reforestation and deforestation shall be decided upon by the first Conference of the Parties serving as Meeting of the Parties (COP/MOP).[31] The indeterminate wording of these provisions led to protracted discussions in the years after Kyoto, and was part of the reason why COP6 failed in 2000.[32] Eventually, the Marrakech Accords resolved many of the outstanding issues regarding sinks and clarified how they should be accounted for.[33] Parties agreed on several principles for implementing LULUCF activities,[34] provided definitions for key terms such as 'afforestation', 'reforestation' and 'deforestation',[35] and clarified the activities that could be accounted for under Article 3.4.[36]

While the Marrakech Accords thus provided a much-needed compromise on the use of forest carbon sinks in achieving the Kyoto targets, another hot potato remained on the table: the use of forestry activities in the CDM. In Marrakech, parties already decided that only afforestation and reforestation, and hence not deforestation, would be eligible under the CDM, and only to a very limited extent.[37] It took two more years to negotiate detailed modalities and procedures to address a host of potential problems related to the inclusion of sinks in the CDM.[38] This included the issue of 'non-permanence', i.e. the risk that forests lose their

---

30 Kyoto Protocol, Art. 3.3.

31 Kyoto Protocol, Art. 3.4.

32 Bailey 2002, 106.

33 Yamin and Depledge 2004, 122–129.

34 Kyoto Protocol Decision 16/CMP.1, para. 1(e).

35 Kyoto Protocol Decision 16/CMP.1, Annex, para. 1.

36 These activities are revegetation, forest management, cropland management, and grazing land management. Kyoto Protocol Decision 16/CMP.1, Annex, para. 6.

37 The cap for credits from forest carbon sinks in the first commitment period is 1%. See Kyoto Protocol Decision 16/CMP.1, Annex, paras. 13–14.

38 Graichen 2005, 11–16.

function as sinks and become sources of carbon emissions if they are harvested or affected by pests, forest fires, etc.[39] Furthermore, there were concerns that forestry activities that resulted in emission reductions in one location would cause an emissions increase elsewhere, nationally and even internationally, also known as 'leakage'. Another issue that needed to be resolved related to accounting for LULUCF emissions and changes in such emissions over time. In particular, there were questions about how to establish whether emission reductions would be 'additional' compared to a business-as-usual scenario.[40] Finally, the inclusion of sinks in the CDM raised concerns about their socio-economic and environmental impacts. Eventually, parties settled their differences and agreed on the inclusion of afforestation and reforestation in the CDM in 2003.[41]

In the context of the negotiations on a post-2012 climate agreement, forests again became a prominent subject of contestation, but this time as a source of emissions. The starting point for these discussions was a proposal by Costa Rica and Papua New Guinea to discuss options for reducing emissions from deforestation in tropical countries at COP11 in 2005.[42] This issue became known as reducing emissions from deforestation and forest degradation (REDD)[43] and was included as an official negotiation item in the 2007 Bali Action Plan.[44] Through REDD, countries with tropical forests could be compensated for their efforts to reduce (the rate of) deforestation and forest degradation.[45] While the idea of creating incentives for reducing deforestation is hardly contested, negotiations have been characterized by disagreement about the specific design of a REDD mechanism, with one of the key questions being

---

39 Graichen 2005, 13–14.

40 Graichen 2005, 14–15.

41 Kyoto Protocol Decision 5/CMP.1. The issue of permanence was resolved by introducing temporary credits under the CDM, whereas the rules also provide safeguards to ensure that real emission reductions take place. See Graichen 2005, 15.

42 UNFCCC 2006a, paras. 76–84.

43 Over time, the name of the agenda item has changed along with the scope of the mechanism. What started as 'RED' (reducing emissions from deforestation in developing countries), became known later as 'REDD' (also including forest degradation) and is now commonly referred to as 'REDD+' (also including the conservation and sustainable management of forests, as well as the enhancement of forest carbon stocks). See Lawlor *et al.* 2010, 5. For the sake of clarity, I use the term REDD consistently to refer to all three, unless stated otherwise.

44 UNFCCC Decision 1/CP.13, para. 1(b)(iii); see also UNFCCC Decision 2/CP.13.

45 Or, as under REDD+, enhance their forest carbon stocks.

whether such a mechanism should be primarily market or fund based, or a combination thereof.[46] Other questions relate to some of the same issues that troubled negotiators in the early 2000s, including monitoring and accounting methods, leakage and non-permanence.[47]

The 2009 Copenhagen Accord recognized the need for the 'immediate establishment' of a REDD mechanism.[48] Furthermore, the meeting adopted a decision on methodological guidance for REDD activities in developing countries.[49] Parties were not yet able to agree on a broader REDD decision, but one year later REDD became an important part of the Cancún Agreements.[50] According to the decision adopted in Cancún, the stated objective for REDD is 'to slow, halt and reverse forest cover and carbon loss'.[51] The fact that the text refers separately to the problems of forest cover loss and carbon loss indicates a shift away from the notion that forests are solely sources or sinks of carbon. The decision calls on developing countries to undertake REDD activities in three phases, starting with national planning, followed by implementation of policies and measures and, eventually, results-based actions.[52] It also requests developing countries to put in place the necessary infrastructure for undertaking REDD activities, including the development of a national strategy, a national reference emission level and a national monitoring system.[53] Finally, the decision contains specific language on social and environmental safeguards that must be respected whilst implementing REDD activities.[54] The decision was an important milestone in the development of REDD in the UN climate regime, but it left many issues unresolved. For instance, parties did not settle on the precise substantive

---

46 Stockwell *et al.* 2009, 151.

47 O'Sullivan 2008, 182–187.

48 UNFCCC Decision 2/CP.15, para. 6. While no REDD mechanism as such is in place under the climate regime, various international REDD activities are paving the way. One of the most notable initiatives is the United Nations Collaborative Programme on Reducing Emissions from Deforestation and Forest Degradation in Developing Countries (UN-REDD), established by the United Nations Development Programme, UNEP and the Food and Agriculture Organization of the United Nations. Other important funding initiatives include the World Bank's Forest Carbon Partnership Facility, and several bilateral partnerships, such as the Norway–Indonesia REDD Partnership.

49 UNFCCC Decision 4/CP.15, para. 6.

50 UNFCCC Decision 1/CP.16.

51 UNFCCC Decision 1/CP.16, Part C, preamble.

52 UNFCCC Decision 1/CP.16, para. 73.

53 UNFCCC Decision 1/CP.16, para. 71.

54 UNFCCC Decision 1/CP.16, para. 72 and Annex I.

scope of the mechanism, as well as its connection to the Green Climate Fund established in Cancún.[55] Modest progress was made at COP17 in Durban, for instance with respect to reporting on safeguards and establishing reference levels,[56] but no consensus could be reached regarding crucial questions on financing REDD. Significant progress was made at COP19 in Warsaw, where parties adopted a series of decisions known as the 'Warsaw Framework for REDD+',[57] which became one of the key achievements of the conference.

In short, throughout the last two decades of climate negotiations, forests have played an important role primarily because of their function as carbon sinks or sources. This role came to the forefront in the aftermath of Kyoto, when the use of sinks for compliance with the Kyoto targets became a crunch issue in the negotiations. The Marrakech Accords were not the end of the story, as the equally vexed question of whether and how to include forestry projects in the CDM was only tackled two years later. Since 2005, forests have also become a key negotiation item following proposals to introduce a REDD mechanism. While the non-mitigation aspects of forests have not been completely ignored, they have only played a subsidiary role. This is in contrast to the more holistic approach taken by the biodiversity regime.

2.3 The Biodiversity Regime and Climate Change

General provisions

The objectives of the CBD are the conservation of biological diversity, the sustainable use of its components, and the fair and equitable sharing of the benefits arising out of the utilization of genetic resources.[58] While the three broad objectives do not make a specific connection to climate change mitigation and/or adaptation, the preamble to the CBD states 'that it is vital to anticipate, prevent and attack the causes of significant reduction or loss of biological diversity at source'.[59] As climate change is one of the major drivers of biodiversity loss, combating climate change thus contributes to achieving the objectives of the biodiversity regime. Other provisions of the CBD are also arguably applicable to climate

---

55 Austin *et al.* 2010.

56 UNFCCC Decision 12/CP.17.

57 The framework consists of seven independent REDD-related decisions, which are available at http://unfccc.int/2860.php#decisions (accessed 11 December 2013).

58 CBD, Art. 1.

59 CBD, preamble.

change. For example, parties are to '[i]dentify processes and categories of activities which have or are likely to have significant adverse impacts on the conservation and sustainable use of biological diversity'.[60] Climate change could be said to be included in these processes.[61]

The biodiversity regime and forests

The CBD is the first multilateral environmental agreement that approaches the protection of biological diversity in a comprehensive fashion, going beyond the piecemeal approaches that characterized international regulatory efforts before its adoption.[62] This holistic consideration of the conservation and sustainable use of biodiversity is reflected in the 'ecosystem approach' adopted by the CBD parties in 1995 and promoted by the Convention since.[63] Among other things, the ecosystem approach embraces community-based ecosystem protection by encouraging decentralization of management to the lowest appropriate level.[64] It also points to the need for considering ecosystems in their economic context, meaning that economic incentives should be used in ecosystem management.[65] Furthermore, interrelations between different ecosystems, for instance within and outside of protected areas, are stressed.[66] The ecosystem approach can be seen as encompassing a variety of approaches to the management and protection of biological resources; it does not prescribe a specific approach, as this will always depend on the prevailing conditions in a certain area.

Whereas the importance of forests under the climate treaties lies in their role as carbon sinks or sources, the CBD is aimed at a range of functions carried out by forests, including the conservation of biological diversity, habitat protection of flora and fauna, the protection of natural heritage and cultural and spiritual values.[67]

The CBD has only slowly expanded its activities in the area of forests, even though parties have repeatedly acknowledged that the issue is covered by the CBD's mandate.[68] While the treaty does not explicitly refer to forests, various provisions are directly or indirectly relevant for

---

[60] CBD, Art. 7(c).
[61] McNeely 2003, 40.
[62] Birnie and Boyle 2002, 568–569.
[63] CBD Decision II/8, para. 1.
[64] CBD Decision V/6, Annex B, para. 6, Principle 2.
[65] CBD Decision V/6, Annex B, para. 6, Principle 4.
[66] CBD Decision V/6, Annex B, para. 6, Principle 3.
[67] Ruis 2001.
[68] Le Prestre 2002, 276.

the protection of forests.[69] Forests, after all, form a part of, and are a habitat of, terrestrial biological diversity.[70] Relevant provisions include obligations with regard to *in situ* and *ex situ* conservation of biodiversity. *In situ* conservation measures include 'the protection of ecosystems, natural habitats and the maintenance of viable populations of species in natural surroundings',[71] as well as the establishment of a system of protected areas or areas where special measures are needed.[72] *Ex situ* conservation measures include the establishment of facilities for research outside of their natural habitat.[73] The CBD also emphasizes the role of indigenous and local communities in the conservation and sustainable use of biodiversity.[74]

The CBD's activities on forests took off in 1996 when the parties requested that the CBD secretariat develop a work programme for forest biodiversity.[75] The initial focus of the work programme was to be on research and the development of technologies relevant for the conservation and sustainable use of forest biodiversity. A first, three-year work programme was subsequently endorsed at COP4 in 1998.[76] In this decision, the CBD secretariat was also asked for the first time to liaise with secretariats of the other Rio Conventions, given 'the potential impact of afforestation, reforestation, forest degradation and deforestation on forest biological diversity and on other ecosystems'.[77]

The scope of the first work programme on forest biodiversity was rather narrow.[78] This was acknowledged four years later, when parties adopted an expanded work programme[79] aimed at more practical action compared to the initial programme.[80] A review carried out by the CBD secretariat deemed the work programme a 'valuable tool', but identified a number of barriers to its effective implementation, including insufficient data, and a lack of capacity.[81] The review drew attention to deforestation and forest degradation as important drivers of forest biodiversity loss, and

---

69 D. Humphreys 2006, 191–204.
70 Khalastchi and Mackenzie 1999, 40.
71 CBD, Art. 8(d).
72 CBD, Art. 8(a).
73 CBD, Art. 9(b).
74 CBD, Art. 8(j).
75 CBD Decision III/12, Annex, para. 6.
76 CBD Decision IV/7, para. 1.
77 CBD Decision IV/7, para. 9.
78 Khalastchi and Mackenzie 1999, 46.
79 CBD Decision VI/22, para. 10 and Annex.
80 D. Humphreys 2006, 192.
81 CBD 2007, 1–2; 13–14.

urged parties to incorporate climate change in their national strategies and programmes.[82] In response to the review, the COP recognized that there is an 'urgent need to strengthen implementation of the programme of work on forest biodiversity … through sustainable forest management and the ecosystem approach'.[83]

The Strategic Plan for Biodiversity 2011–2020 (and its Aichi targets), adopted at COP10 in Nagoya in 2010, contains several non-legally binding targets related to forest biodiversity. For instance, it states that, by 2020: the rate of loss of natural habitats, including forests, needs to be at least halved (and where feasible brought to zero);[84] areas under forestry need to be managed sustainably;[85] and the contribution of biodiversity to carbon stocks needs to be enhanced, including at least 15 per cent of degraded ecosystems.[86]

Overall, the depth and scope of the CBD's activities in the area of forest biodiversity have significantly increased, especially after the launch of the expanded work programme. The ecosystem approach offers general guidance for the CBD's activities and the CBD's view of forests is generally broader in scope than that of the climate regime. This means, for example, that the CBD considers how ecosystems relate to each other. Furthermore, this brief review shows that the CBD parties have drawn attention to the forest-related linkages between climate change and biodiversity.

3. CONSEQUENCES OF INTERACTIONS

This section shows how decisions taken in the climate regime might have negative impacts on achieving the objectives of the CBD. At the same time, it stresses that there is an untapped potential for synergies between the regimes on the issue of forests, particularly in light of the emergence of REDD.[87]

---

82 CBD 2007, 2.

83 CBD Decision IX/5, preamble.

84 CBD Decision X/2, Annex, Target 5.

85 CBD Decision X/2, Annex, Target 7.

86 CBD Decision X/2, Annex, Target 15.

87 I focus on the climate regime as the source of interactions and the biodiversity regime as the target. Interactions may also go in the reverse direction. For instance, CBD parties in 2010 agreed on a moratorium on geoengineering, thereby seemingly limiting a potential climate change mitigation option (albeit a controversial one). See CBD Decision X/33, para. 8(w).

3.1 Forest Carbon Sinks in the Kyoto Protocol

There are concerns that the use of sinks in the Kyoto Protocol leads to conflicts between the climate and biodiversity treaties in the implementation phase.[88] More specifically, critics have argued that the rules developed under the Kyoto Protocol do not sufficiently safeguard biodiversity concerns and could frustrate the objectives of the biodiversity treaty.[89] The main concerns raised are that Kyoto's rules could result in destructive large-scale, monoculture plantations, a lack of protection for existing old-growth forests and an increase in the use of genetically modified trees and invasive alien species.

The impacts of forest-related mitigation activities on biodiversity may vary, depending on the type and the design of such activities. Some activities, such as the preservation of natural forests through the prevention of deforestation and forest degradation, can result in significant biodiversity benefits.[90] By contrast, using the wrong sites or species for afforestation and reforestation may lead to negative effects on biodiversity.[91] One of the main concerns is the replacement of old-growth forests by large-scale industrial plantations, which use fast-growing trees that sequester more carbon.[92] While the mitigation benefits are clear, particularly in the short term, the impacts on biodiversity associated with plantations can be negative. In particular monoculture plantations, which use only one tree species, can lead to adverse effects on plant and animal biodiversity.[93] The detrimental effects on biodiversity are aggravated if the plantation consists of genetically modified trees. There are risks that genetically modified trees might be, or might create, invasive species that threaten other species in an ecosystem.[94]

The policy conflict between the Kyoto Protocol and the CBD is thus shaped by the type of incentives provided by the rules on forest carbon

---

However, in the case of forests the direction of interactions is primarily from the climate regime to the biodiversity regime.

[88] Pontecorvo 1999; Jacquemont and Caparrós 2002; Wolfrum and Matz 2003, 79–93; Sagemüller 2006; Schwartz 2006.

[89] Pontecorvo 1999, 749.

[90] Secretariat of the CBD 2009a, 102.

[91] Sagemüller 2006, 197–198.

[92] Pontecorvo 1999, 49; Sagemüller 2006, 199.

[93] Jacquemont and Caparrós 2002, 174. However, using various native species would reduce the impacts of plantations. See Secretariat of the CBD 2009a, 104. Furthermore, even monoculture plantations may lead to biodiversity benefits. See Sagemüller 2006, 199.

[94] Schwartz 2006, 435–436.

sinks developed under the Protocol. The outcomes of this relationship depend on whether these rules only seek to maximize the carbon sequestration potential of forests or whether they limit such behaviour given the associated potential biodiversity impacts. The decisions on LULUCF and on the inclusion of afforestation and reforestation in the CDM provide various safeguards for biodiversity protection, but these incentives are rather weak.

First, the principles for the implementation of LULUCF activities include ensuring '[t]hat the implementation ... contributes to the conservation of biodiversity and sustainable use of natural resources'.[95] However, the use of the weak phrasing 'contributes to' arguably does not result in any concrete limitations of activities for the benefit of biodiversity.[96]

A second way in which biodiversity considerations could limit LULUCF activities is through the definition of 'forest management'. Forest management includes activities such as harvesting, forest fire management, pest management and regeneration, all of which may impact biodiversity.[97] In the Marrakech Accords, forest management is defined as 'a system of practices for stewardship and use of forest land aimed at fulfilling relevant ecological (including biological diversity), economic and social functions of the forest in a sustainable manner'.[98] This formulation again lacks specificity. It is unclear, for example, what is meant by 'stewardship' or what 'relevant' ecological functions comprise.[99] Moreover, the reference to biodiversity is found only in the definition of 'forest management', and not in the definitions of 'afforestation', 'reforestation' or 'deforestation', risking 'the dangerous interpretation that biodiversity conservation is only necessary in the case of forest management'.[100] Despite these shortcomings, the definition seems to at least preclude forest management activities that have clear adverse impacts on biodiversity.[101]

Third, afforestation and reforestation projects in the CDM require an analysis of the socio-economic and environmental impacts, including impacts on biodiversity and natural ecosystems.[102] If such an analysis

---

95 Kyoto Protocol Decision 16/CMP.1, para. 1(e).
96 Jacquemont and Caparrós 2002, 171.
97 Secretariat of the CBD 2003, 65.
98 Kyoto Protocol Decision 16/CMP.1, para. 1(f).
99 Sagemüller 2006, 210–211.
100 Jacquemont and Caparrós 2002, 171–172.
101 Sagemüller 2006, 211–212.
102 Kyoto Protocol Decision 5/CMP.1, Annex, para. 12(c).

points to significant impacts, an environmental impact assessment is necessary. In theory, the requirement to conduct an environmental impact assessment is another safeguard mechanism for biodiversity protection. In practice, the usefulness of this mechanism is again constrained. The determination of which impacts are 'significant' is left up to the host country and project participants.[103] This is problematic because participants have an economic incentive to abstain from conducting an impact assessment to keep project costs down,[104] whereas CDM host countries have incentives to attract CDM investments. The decision also states that environmental impact assessments need to be carried out 'in accordance with the procedures required by the host Party',[105] but fails to provide any guidance on how parties should set up an environmental impact assessment.[106] Moreover, the decision also does not account for impacts that might only manifest themselves over time.[107] Nevertheless, the impact assessment should allow for the identification, and possibly mitigation, of biodiversity impacts in some cases, and result in a basic level of transparency concerning the environmental impacts of CDM sink projects.[108]

The fourth reference to biodiversity considerations in the Marrakech Accords is that parties, in their national reports, should describe the laws and administrative procedures in place to ensure that LULUCF activities contribute to biodiversity conservation and the sustainable use of natural resources.[109] While reporting is an important part of promoting compliance, there are no consequences attached to inaccurate or incomplete reporting on this aspect.[110]

The fifth limitation is specifically aimed at preventing the conversion of natural, old-growth forests into plantations. The rules seek to achieve this by limiting 'afforestation' to 'land that has not been forested for a period of at least 50 years' and 'reforestation' to 'those lands that did not contain forest on 31 December 1989'.[111] These provisions provide one of

---

103 Meinshausen and Hare 2003, 6.
104 Sagemüller 2006, 213.
105 Kyoto Protocol Decision 5/CMP.1, Annex, para. 12(c).
106 Sagemüller 2006, 213.
107 Meinshausen and Hare 2003, 6.
108 Sagemüller 2006, 212.
109 Kyoto Protocol Decision 15/CMP.1, para. 38.
110 Jacquemont and Caparrós 2002, 172.
111 Kyoto Protocol Decision 16/CMP.1, paras. 1(b) and (c).

the most important general safeguards, even though they are not explicitly linked to biodiversity protection as such. However, some commentators have warned that there is still a risk of conversion through forest management practices.[112]

Finally, the Marrakech Accords include a general statement instructing CDM host countries to evaluate the risks associated with the use of genetically modified organisms and potentially invasive alien species, and instructing developed countries to evaluate the use of CDM credits resulting from project activities that make use of them.[113] Even if this statement could be interpreted as an obligation to evaluate (which is already questionable), it could only be operationalized if parties knew whether potentially invasive alien species or genetically modified organisms are being used, and the rules on this point are not entirely clear.[114] Moreover, given that projects using genetically modified trees tend to be cost efficient,[115] there are fewer incentives for a host country to block a project proposal.[116]

There is thus an actual policy conflict between the Kyoto Protocol and CBD, primarily due to the relatively strong incentives under the Kyoto Protocol to achieve emission reductions cost effectively, be it through the accounting of LULUCF activities for the purposes of achieving the Kyoto targets or through the CDM. In contrast, the incentives to protect biodiversity while implementing such activities are procedural or formulated broadly. The main exception lies in the definitions of 'afforestation' and 'reforestation', which reduce the incentive to convert natural forests into plantations.

3.2 Tackling Deforestation

The preceding discussion shows that rules providing the 'wrong' incentives have resulted in a policy conflict between the climate and biodiversity regimes. However, it may have painted too bleak a picture of the outcomes of interactions between the climate and biodiversity regimes. The extent of the conflict should not be exaggerated. For instance, the actual practice shows that forestry projects in the CDM never really took off, mainly due to financial constraints (e.g., high transaction costs), a

---

112 Jacquemont and Caparrós 2002, 172; Sagemüller 2006, 214–215.
113 UNFCCC Decision 19/CP.9, preamble.
114 Meinshausen and Hare 2003, 5.
115 Schwartz 2006, 469.
116 Sagemüller 2006, 216.

lack of knowledge and capacity in host countries and their exclusion from the EU's emissions trading system.[117]

More importantly, the emergence of REDD has presented a clear opportunity to capture synergies between the climate and biodiversity treaties, given the large benefits REDD may entail for both biodiversity protection and climate change mitigation.[118] Given the failure of the international community to provide for adequate protection of the world's forests, hopes have been raised that the biodiversity regime could possibly hitch a ride with the climate regime at a time where climate change is generally higher on the agenda of policymakers than the protection of biodiversity.[119] Drawing attention to the overlapping issues could lead to prioritizing (and possibly also to funding of) climate change activities with positive spillovers for biodiversity.

As noted above, the contours of a REDD mechanism are slowly emerging. The precise impacts on biodiversity will depend on the design details as well as REDD's implementation in practice. What, then, are the design features of a REDD mechanism that matter from a biodiversity perspective? First, the scope is important: REDD, as well as forest conservation,[120] would generally lead to greater biodiversity benefits than the enhancement of carbon stocks and sustainable management of forests.[121] Related to the scope, it matters whether the definition of 'forests' under REDD distinguishes between natural forests and plantation forests.[122] In particular, how forests are defined could determine whether a REDD mechanism provides incentives to convert natural forests into plantations. Second, how a REDD mechanism tackles the issue of leakage is relevant for biodiversity protection, especially if the prevention of deforestation in areas with low levels of biodiversity leads to an increase in deforestation in other areas with high levels of biodiversity.[123] Third, the use of a market-based approach to REDD (as opposed to a fund) is more likely to steer investments to areas with high carbon emissions, which do not necessarily have the highest level of

---

117 Thomas *et al.* 2010, 881–884.
118 Secretariat of the CBD 2009a, 102; Harvey *et al.* 2010, 53.
119 On this idea of 'bandwagoning', see Jinnah and Muñoz Cabré 2011.
120 The first element of 'REDD+'.
121 The other two elements of 'REDD+'. See also Harvey *et al.* 2010, 54–55.
122 Harvey *et al.* 2010, 55.
123 Karousakis 2009, 13; Harvey *et al.* 2010, 55.

biodiversity.[124] Finally, the inclusion of monitoring obligations covering the biodiversity impacts of REDD activities could indirectly support biodiversity objectives by making the way REDD affects biodiversity more transparent.[125] While parties' positions on the details of these design features diverge, there is consensus on some basic elements. For example, parties agree that REDD should encompass forest conservation, the sustainable management of forests, and the enhancement of carbon stocks, even though the initial focus of REDD in practice may be on avoiding deforestation and forest degradation.[126] Furthermore, it is clear that the monitoring of safeguards primarily takes place at the national level, through 'systems for providing information on how safeguards are addressed and respected'.[127]

The decisions on REDD adopted under the UNFCCC provide an indication about how the climate regime will accommodate biodiversity objectives in REDD design.[128] In the first REDD decision, parties provided indicative guidance for REDD demonstration activities, indicating that such activities 'should be consistent with sustainable forest management, noting, inter alia, the relevant provisions of … the Convention on Biological Diversity'.[129] In the REDD decision adopted in Copenhagen, parties noted 'the importance of promoting sustainable management of forests and co-benefits, including biodiversity, that may complement the aims and objectives of national forest programmes and relevant international conventions and agreements'.[130] The Cancún Agreements contain a more specific formulation, indicating that parties should promote and support safeguards. With respect to biodiversity, this means that

> [a]ctions [need to be] consistent with the conservation of natural forests and biological diversity, ensuring that [REDD activities] are not used for the conversion of natural forests, but are instead used to incentivize the protection and conservation of natural forests and their ecosystem services, and to enhance other social and environmental benefits.[131]

---

[124] Brown *et al.* 2008, 117; Venter *et al.* 2009, 1368. However, a drawback of a fund-based approach is that it may lead to significantly lower levels of funding. See Brown *et al.* 2008, 117.

[125] Savaresi 2011, 99–101.

[126] Verchot and Petkova 2009, 8.

[127] UNFCCC Decision 12/CP.17.

[128] Pistorius *et al.* 2010, 3.

[129] UNFCCC Decision 2/CP.13, Annex, para. 8.

[130] UNFCCC Decision 4/CP.15, preamble.

[131] UNFCCC Decision 1/CP.16, Appendix I, para. 2(e).

Another safeguard requires parties to ensure that '[a]ctions complement or are consistent with the objectives of … relevant international conventions and agreements'.[132] Given the clear overlap, there is a strong argument that the CBD is 'relevant' for the purposes of this decision. The decision also outlines several general principles for REDD activities. While not all of these are relevant from a biodiversity perspective, one of them states that REDD activities should '[b]e consistent with the objective of environmental integrity and take into account the multiple functions of forests and other ecosystems'.[133] This phrasing seemingly confirms the shift away from the idea that forests are merely sources or sinks of carbon. The REDD decision adopted in Durban does not contain provisions related to biodiversity safeguards, but instructs parties implementing REDD to enhance transparency on how safeguards are addressed and respected.[134] Parties need to do so through their National Communications to the UNFCCC or, on a voluntary basis, through a web-based platform.[135]

The decisions show that parties pay attention to the potential impacts of REDD on biodiversity, but that they have mainly dealt with these impacts in terms of 'co-benefits' or 'safeguards', implying that the primary objective is to maximize the reduction of CO_2 emissions. For some observers, the inclusion of biodiversity considerations in the design of REDD indeed poses a barrier to the mechanism's implementation.[136] The co-benefits approach is one way to integrate biodiversity issues in REDD. Yet biodiversity concerns could also be seen as more than just auxiliary benefits and rather as separate objectives for a REDD mechanism for which complementary funding mechanisms would be required. Or, going one step further, REDD should arguably deliver biodiversity benefits even if this might be at the expense of its mitigation goals.[137]

Looking back, the case of tackling deforestation points to a potential policy conflict but also to potential synergy between the climate regime and the CBD. More generally, two narratives summarizing the relationship between the climate and biodiversity treaties with respect to forests are discernable. The narrative of conflict highlights and criticizes the ambiguities and lack of safeguards in the relevant rules developed in the climate regime. Similarly, the narrative of synergies emphasizes the

---

132 UNFCCC Decision 1/CP.16, Appendix I, para. 2(a).
133 UNFCCC Decision 1/CP.16, Appendix I, para. 1(d).
134 UNFCCC Decision 12/CP.17, para. 2.
135 UNFCCC Decision -/CP.19, paras. 3–4.
136 E.g., CBD 2010a, 16.
137 Harvey *et al.* 2010, 57–59.

potential co-benefits that could be achieved through implementing forestry activities, in particular through the establishment of a REDD mechanism. These narratives are not mutually exclusive but serve to illustrate that the relationship between the two regimes can be both conflictive and synergistic, depending on how actors within and outside the regimes choose to manage the interactions.

4. INTERACTION MANAGEMENT: LEGAL TECHNIQUES

4.1 Conflict Avoidance Techniques

Treaty changes and drafting

The first way to avoid a potential conflict between the climate and biodiversity treaties is by amending the existing treaties. The ongoing climate negotiations present an opportunity to do so. Various proposals on how to include biodiversity concerns in rules developed in the climate regime have been put forward by commentators.[138] The most far-reaching would likely be the establishment of legally binding international biodiversity standards.[139] However, such standards would likely meet with the same developing country opposition that blocked the use of sustainable development criteria under the CDM.[140] Another measure is the exclusion of specific forestry activities that may have negative biodiversity impacts from the scope of REDD.[141] Other options include the international requirement of mandatory environmental impact assessments for REDD projects, or more broadly requiring the use of strategic environmental assessments.[142] Furthermore, monitoring and reporting provisions that cover biodiversity impacts of REDD projects could make the relationship between the climate change and biodiversity regimes more transparent.

While a host of options is thus available to parties, it is doubtful that their adoption is politically feasible. The slow-moving discussion on including biodiversity considerations in a REDD mechanism provides an indication of the political sensitivities. Moreover, even if negotiators

---

[138] Sagemüller 2006, 224–232; Ebeling and Fehse 2009, 35–40; Pistorius *et al.* 2010, 19–23; van Asselt 2011a, 144–147.

[139] Sagemüller 2006, 224–225; Ebeling and Fehse 2009, 35–36.

[140] Ebeling and Fehse 2009, 36.

[141] Ebeling and Fehse 2009, 38–39.

[142] Sagemüller 2006, 225–230.

make use of this window of opportunity, it is more likely that they will do so by addressing biodiversity issues through a COP decision, rather than through amending one of the climate treaties or as part of a new climate treaty. Treaty body decisions are generally associated with greater flexibility and are usually easier to negotiate than legally binding treaties.[143] Furthermore, there is a precedent in how biodiversity considerations have been integrated in the LULUCF rules under the Kyoto Protocol. It may be possible to agree on a brief reference to the contribution of REDD to biodiversity benefits in a legally binding agreement, a provision which could subsequently be operationalized through treaty body decisions.[144]

Treaty interpretation

It is difficult to envisage a dispute relating to the climate–biodiversity overlap being brought before an international court or arbitrator. However, treaty interpretation may also be used by government officials, legal advisers and domestic courts.[145] Alternatively, interpretation may take place if parties jointly agree on an authentic interpretation of provisions in the climate treaties taking into account the CBD.[146] Such an activity resembles an amendment[147] and will hence be limited by the political considerations noted above.[148]

Nevertheless, treaty interpretation has the potential to harmonize seemingly inconsistent norms. Given the overlaps between the objectives of the climate and biodiversity treaties, a teleological interpretation could support the harmonization of the treaties. The utility of teleological interpretation should not be exaggerated though: the wording of the objectives of the climate and biodiversity treaties is very general and does not necessarily help to clarify the meaning of the specific provisions on sinks.[149]

Article 31.3(c) of the VCLT may also play a role. For Concetta Maria Pontecorvo, this provision confirms 'a specific duty for Parties to interpret the provisions of the Kyoto Protocol relating to sinks potentially

---

143 Raustiala 2005, 591–592.

144 van Asselt 2011a, 148.

145 Gardiner 2008, 110; Matz-Lück 2008b, 50–52; Matz-Lück 2012, 213–218.

146 Cf. VCLT, Art. 31.3(a).

147 Gardiner 2008, 110.

148 Wolfrum and Matz 2003, 140.

149 Wolfrum and Matz 2003, 135–137.

conflicting with pre-existing commitments under other treaties in such a way as to make them compatible with these commitments'.[150]

More generally, an evolutionary approach to treaty interpretation is relevant for the climate and biodiversity treaties. For instance, the term 'sustainable forest management practices'[151] is prone to changing scientific insights. Some parties have followed this approach in the case of the climate and biodiversity treaties. For instance, a submission by the Alliance of Small Island States argues that '[t]he implementation of activities under Article 3.3 and 3.4 of the Kyoto Protocol must respect the requirements set out in the [CBD and UN Convention to Combat Desertification]'.[152]

While there is potential for using treaty interpretation as a technique to avoid conflicts between the climate and biodiversity treaties, the general limitations to its use should be remembered (see Chapter 5). Nevertheless, treaty interpretation by domestic actors provides an important avenue to pursue a 'harmonizing approach' in the implementation of the two treaties when there is no agreement on a specific treaty term at the international level (see Section 6 below).[153]

4.2 Conflict Resolution Techniques

Conflict clauses

With respect to conflict resolution techniques, the first question is whether the climate and biodiversity treaties contain any conflict clauses. The Kyoto Protocol's reference to 'relevant international environmental agreements',[154] which instructs parties to consider their commitments under other treaties, arguably constitutes such a conflict clause. Although this provision does not state *which* agreements need to be taken into account, it is reasonable to assume that, given the role of forests as sinks and sources of emissions on the one hand and as part and habitat of biodiversity on the other, the CBD can be considered 'relevant'.[155] Still, the provision is unclear about which commitments in other agreements it refers to, and also merely states that such commitments should be 'taken into account' (rather than suggesting that they need to be complied

---

150 Pontecorvo 1999, 741.
151 Kyoto Protocol, Art. 2.1(a)(ii).
152 UNFCCC 2001a, 4.
153 Wolfrum and Matz 2003, 146.
154 Kyoto Protocol, Art. 2.1(a)(ii).
155 Sagemüller 2006, 208; see also Depledge 2000, para. 87.

with).[156] It is therefore difficult to see how this formulation regulates how the commitments in the Kyoto Protocol are related to those contained in the CBD.[157]

The CBD contains a more clearly identifiable conflict clause. The provision gives priority to any existing agreement 'except where the exercise of those rights and obligations would cause a serious damage or threat to biological diversity'.[158] In effect, this formulation reverses the conflict clause:[159] the clause serves to limit climate change mitigation activities that would cause a serious damage or threat to biodiversity. However, several caveats apply. The clause applies only to treaties *existing at the time* of the CBD's adoption, and is thus not applicable to the subsequently adopted Kyoto Protocol. This makes this avenue for addressing the potential conflict between the two treaties 'legally minimal and practically nonexistent'.[160] Furthermore, the phrase 'serious damage or threat to biological diversity' is nowhere defined or elaborated upon, meaning that the practical application of the clause remains uncertain and states have a rather wide margin of appreciation. While the notion contained in Article 22.1 may contain 'a rule of precaution and proportionality, in which a hierarchy between regimes is established on a case-by-case basis',[161] there is no clear guidance for any adjudicator to apply such a rule. Hence, I can only concur with Rüdiger Wolfrum and Nele Matz, who conclude that '[i]t is doubtful that this clause can prevent or solve conflicts'.[162]

Priority rules

The conflict resolution rules of *lex posterior* and *lex specialis* may also apply. Article 30 of the VCLT only applies to 'successive treaties relating to the same subject matter'. If this subject matter is broadly seen as 'environmental protection', the Vienna Convention could apply to conflicts between the climate and biodiversity treaties. Alternatively, 'if two different rules or sets of rules are invoked in regard to the same matter',[163] the conflict resolution rules can arguably be applied. However,

---

156 Pontecorvo 1999, 739–740.
157 Jacquemont and Caparrós 2002, 178.
158 CBD, Art. 22.1.
159 Wolfrum and Matz 2003, 124.
160 Doelle 2004, 86.
161 Werksman 1999a, 5.
162 Wolfrum and Matz 2003, 125; see also Fitzmaurice and Elias 2005, 344–345; ILC 2006, para. 280.
163 ILC 2006, paras. 23 and 254.

such a broad interpretation would be inappropriate[164] as it would make the very existence of a conflict entirely dependent on the rules invoked by parties involved in a conflict.

Assuming that the two treaties relate to the same subject matter, it is still not straightforward to establish which treaty is later in time.[165] Even if it would be possible to agree on the 'date' of a treaty (e.g., the time of its adoption, the date of entry into force or the date of ratification or accession by a party), the use of the *lex posterior* rule is not appropriate in the case of treaties that were negotiated in parallel, as is the case for the UNFCCC and the CBD.[166]

Moreover, application of the conflict rules on successive treaties is difficult to envisage in the case of 'living treaties', where 'treaty norms are part of a regulatory framework or legal system that was created at one point in time but continues to exist and evolve over a mostly indefinite period'.[167] MEAs are clear examples of such 'living treaties', where parties continue to develop the treaty through decisions by treaty bodies. In such cases, the applicability of the law of treaties is inherently limited. As shown in the previous section, the interactions between the climate and biodiversity regimes with respect to forests are triggered by treaty body decisions, not by the treaties as such.

The applicability of the *lex specialis* maxim is also questionable. Again, even on the assumption that two treaties have the same subject matter they deal with very different aspects of an overlapping issue.[168] As Ronnie Yearwood argues, '[e]ach system is *lex specialis* as to the next system because the purpose of each system is to regulate a specific concern of international law'.[169] From the perspective of the CBD, its own rules would be more specific for protecting forest biodiversity. From the perspective of the climate treaties, the rules developed under them would be more specific in terms of tackling emissions from forests. In other words, the determination of which norm is more specific in terms of subject matter is inevitably in the eye of the beholder.[170]

164 Wolfrum and Matz 2003, 151.
165 Cf. Vierdag 1988; Pauwelyn 2003a, 367–384.
166 Wolfrum and Matz 2003, 155.
167 Pauwelyn 2003a, 378.
168 Lindroos 2005, 48.
169 Yearwood 2011, 57.
170 Wolfrum and Matz 2003, 157–158.

5. INTERACTION MANAGEMENT: INSTITUTIONAL COORDINATION

5.1 Existing Institutional Coordination

Conferences of the Parties

Given that many of the existing interactions between the climate and biodiversity regimes stem from treaty body decisions, it makes sense to examine how the COPs have sought to respond to these interactions. From the various decisions addressing the mutual relationship between the two regimes it appears that while the treaty bodies of the UNFCCC and the Kyoto Protocol have been rather passive on the relationship with the Biodiversity Convention, the CBD COP has actively sought to influence the outcomes of regime interactions.

The story of the climate COP decisions on promoting cooperation with the CBD is a relatively short one, which may be surprising given the potential for synergies and conflicts. Both the UNFCCC COP and the Kyoto Protocol COP/MOP are mandated to '[s]eek and utilize, where appropriate, the services and cooperation of, and information provided by, competent international organizations and intergovernmental and non-governmental bodies',[171] but the initial activities were only in response to calls by the CBD COP. Cooperation with the CBD received its first mention in a 2000 report by the UNFCCC's Subsidiary Body for Scientific and Technological Advice[172] following a note by the CBD's Executive Secretary.[173] Since then, cooperation with the CBD has been discussed under the agenda item of 'cooperation with relevant international organizations'. Only one (brief) COP decision on cooperation has been adopted, which generally affirms the need for enhanced cooperation 'with the aim of ensuring the environmental integrity of the [Rio Conventions] and promoting synergies under the common objective of sustainable development, in order to avoid duplication of efforts, strengthen joint efforts and use available resources more efficiently'.[174] This decision also formally endorses a Joint Liaison Group between the secretariats of the Rio Conventions (see below).[175] Since the adoption of this decision, however, there have been no major efforts to manage

[171] UNFCCC, Art. 7.2(l); Kyoto Protocol, Art. 13.4(i).
[172] UNFCCC 2000, para. 60.
[173] CBD 2000, Annex.
[174] UNFCCC Decision 13/CP.8, preamble.
[175] UNFCCC Decision 13/CP.8, para. 3.

interactions between the two treaties, and the UNFCCC COP's attitude remains reactive.

The CBD parties, in contrast, have adopted various decisions on biodiversity and climate change with a view to avoiding (and responding to) conflicts and promoting synergies on various biodiversity-related issues.[176] These decisions may have been instrumental in highlighting biodiversity concerns in UNFCCC decisions,[177] but they have not led to strong references to biodiversity in the climate regime's decisions on forests. The first decisions were adopted in 2000, when the discussion on sinks in the climate regime was high on the agenda. One of these decisions 'urged' parties to the UNFCCC 'to ensure that future activities of the [UNFCCC], including forest and carbon sequestration, are consistent with and supportive of the conservation and sustainable use of biological diversity', and asked the CBD's Subsidiary Body on Scientific, Technical and Technological Advice to provide scientific advice on how to integrate biodiversity considerations in the implementation of the climate treaties.[178]

The first separate decision on biodiversity and climate change was adopted in 2004, arguing that there are opportunities for synergies.[179] The decision points to a CBD-specific approach to addressing the interactions between the CBD and the climate treaties, noting that the ecosystem approach 'could facilitate the formulation of climate change mitigation and adaptation projects that also contribute to biodiversity conservation and sustainable use at the national level'.[180] The decision further requests the subsidiary body to develop further guidance for promoting synergies, and invites the UNFCCC to collaborate to this end.[181]

Another decision on biodiversity and climate change, adopted in 2006, calls on the CBD parties and other countries to integrate biodiversity considerations into their climate policies.[182] Parties at COP9 in 2008 decided that climate change considerations should be integrated in future work programmes, taking into account, among other things, the ecosystem approach.[183] The decision also invited the UNFCCC to 'take full

---

[176] CBD, Art. 23.4(h); see also Art. 24.1(d).
[177] Yamin and Depledge 2004, 523–524.
[178] CBD Decision V/4, para. 18.
[179] CBD Decision VII/15, para. 7.
[180] CBD Decision VII/15, para. 8.
[181] CBD Decision VII/15, paras. 14–15.
[182] CBD Decision VIII/30, paras. 1–2
[183] CBD Decision IX/16, para. A.1(h).

account of opportunities for its work to provide benefits for biodiversity'.[184] Parties were further invited to implement various activities to promote synergies among the Rio Conventions.[185]

The attention on the climate and biodiversity interactions in the CBD reached its pinnacle at the CBD COP held in October 2010 in Nagoya, Japan.[186] This was in part because a review of the work on biodiversity and climate change was due,[187] but likely also because of the ongoing discussions on REDD in the climate negotiations. The parties adopted a lengthy decision on biodiversity and climate change, which contains some novel elements, as well as proposals that link the CBD more strongly to the REDD discussions under the UNFCCC. For instance, the decision suggested that parties should consider streamlined reporting with respect to overlapping issues.[188] Furthermore, it invited parties to consider specific guidance related to ecosystem-based approaches for climate change mitigation and adaptation.[189] More specifically, it provided several carefully formulated recommendations with a view to minimizing the biodiversity impacts of forestry activities.[190] The decision also requested the CBD secretariat to collaborate with other international bureaucracies and with CBD parties to contribute to the discussion on, and possible development of, biodiversity safeguards.[191] The CBD secretariat was also asked to identify possible indicators for monitoring how REDD could contribute to the objectives of the CBD (see below).[192] Lastly, the decision proposed developing joint activities with the other Rio Conventions.[193]

While the recommendation to streamline reporting is aimed at improving operational inefficiencies, other suggestions can be viewed as efforts to influence the design of climate change measures at the international

---

184 CBD Decision IX/16, para. B.11(b).

185 CBD Decision IX/16, para. B.8, and Annex II. These activities include collaboration among national focal points, cooperation on national-level planning and cooperation in forest sector planning.

186 Morgera 2011.

187 CBD 2010b, para. 1.

188 CBD Decision X/33, para. 7.

189 CBD Decision X/33, para. 8.

190 CBD Decision X/33, para. 8(p).

191 CBD Decision X/33, para. 9(g).

192 CBD Decision X/33, para. 9(h).

193 CBD Decision X/33, para. 13(a); see also CBD Decision X/20, para. 6(a). The CBD's Subsidiary Body for Scientific, Technical and Technological Advice had originally proposed a more far-reaching joint work programme on the basis of a suggestion by the CBD secretariat. See CBD 2010c.

and national levels. These suggestions may have had an effect at the international level, given the references to biodiversity in the Cancún Agreements, although the short time between the two COPs would suggest that there was little opportunity for the Nagoya decision to influence the Cancún outcomes.

The CBD COP in Hyderabad in 2012 sought to maintain momentum, but parties were only able to agree to 'take note with appreciation' of an annex to a decision, which provides more detailed guidance on how to safeguard biodiversity in the implementation of REDD activities at the national level.[194] The guidance follows a CBD secretariat note providing advice on REDD safeguards (see below).

It remains to be seen whether and how the CBD will actually become involved in the operationalization of the biodiversity safeguards under the climate regime and, if so, how effective actions under the CBD would be in mitigating conflicts between the two regimes. One possibility is that it will play a role in monitoring the biodiversity impacts of REDD. While the Cancún Agreements do not preclude this possibility, they also do not specify a role for the CBD.[195]

Secretariats

Liaising with other secretariats is one of the tasks assigned to the secretariats of environmental treaties, including the UNFCCC secretariat[196] and the CBD secretariat.[197] In particular, the CBD secretariat has played an important independent role in terms of knowledge generation and raising awareness about the (potential) interactions, even though most of its activities have been triggered by COP decisions. Through 'marketing' the linkages between the CBD and the climate regime, the CBD secretariat has sought to manage the interactions between the two regimes, straddling the line between 'cognitive' and 'normative' influence.[198]

One of the main developments that involved the bureaucracies of the climate and biodiversity regimes was the creation of the Joint Liaison Group at the request of the CBD's Subsidiary Body for Scientific,

---

194 CBD Decision XI/19, para. 9 and Annex.

195 UNFCCC Decision 1/CP.16, para. 71(d). Further details were agreed upon in Durban, but the decision is still silent about a possible role for the CBD. See UNFCCC Decision 12/CP.17, paras. 1–6.

196 UNFCCC, Art. 8.2(e); Kyoto Protocol, Art. 14.2.

197 CBD, Art. 24.1(d).

198 Jinnah 2011, 26.

Technical and Technological Advice,[199] which was in turn inspired by a resolution of the UN General Assembly (see Chapter 5).[200] The Group brought together the secretariats of the CBD and the UNFCCC, who were later joined by the secretariat of the UN Convention to Combat Desertification. Its mandate is to 'enhance coordination between the three conventions, including the exchange of relevant information' and '[t]o explore options for further cooperation between the three conventions, including the possibility of a joint work plan and/or a workshop'.[201] The emphasis of its activities has been on information exchange, with a view to facilitating the mutually supportive implementation of the three agreements at the national level.[202]

By 2011 the Joint Liaison Group had convened 11 times, focusing on cross-cutting issues such as research and monitoring, information exchange, technology transfer, capacity building, financial resources, education and public awareness, and adaptation to climate change. In 2004 the three secretariats drafted a joint paper identifying options for enhanced cooperation.[203] Whereas some of the options identified in the paper (e.g., joint workshops and the sharing of information among secretariat staff) are relatively easy to implement, others (e.g., the harmonization of reporting) require much more preparation and consensus (see below).

In addition to its activities as a member of the Joint Liaison Group, the CBD secretariat has also individually undertaken various activities delegated to it by the CBD COP.[204] For instance, it has sought to integrate climate change considerations in the various work programmes developed under the CBD.[205] Furthermore, the CBD secretariat has been involved in the work of several ad hoc technical expert groups established to provide scientific and technical advice on issues at the intersection of climate change and biodiversity.[206] Finally, the decision on biodiversity and climate change adopted in Nagoya puts the CBD secretariat in charge

---

199 CBD 2001, para. 9.
200 UNGA 1999.
201 UNFCCC 2001b, para. 42(d); CBD 2001, para. 9.
202 As was underlined by the Executive Secretary of the UNFCCC, Christiana Figueres. See Joint Liaison Group 2011, 1–2.
203 CBD 2004.
204 CBD 2010b, paras. 6–16.
205 CBD 2010b, para. 11.
206 Secretariat of the CBD 2003; Secretariat of the CBD 2009a.

of various important issues related to the interactions with the climate regime, including the development of a proposal for joint activities.[207]

One of these activities in response to a CBD COP decision resulted in a document providing advice on REDD safeguards as well as indicators for biodiversity impacts on REDD. In the document, the CBD secretariat highlighted existing environmental and social safeguards developed through other initiatives, including the World Bank's Forest Carbon Partnership Facility, UN-REDD and private REDD social and environmental standards developed by the Climate, Community and Biodiversity Alliance and CARE International. The secretariat recommended that countries undertaking REDD activities apply these safeguards (individually or in combination) and suggested harmonizing the safeguards.[208] In the same document, the CBD secretariat also provided an overview of how the existing indicators for achieving biodiversity targets[209] could be used as indicators for the monitoring of REDD's biodiversity impacts.[210] However, the secretariat made clear that it did not intend to pre-empt the work on reference levels and safeguards for REDD under the UNFCCC, and suggested only considering modalities for monitoring the biodiversity impacts of REDD when the discussions in the climate regime had 'progressed further'.[211] Nevertheless, the submission of a report of its work on REDD and biodiversity to the UNFCCC COP in Durban in 2011 shows that the CBD secretariat seeks to have the viewpoints from the biodiversity community included in the climate negotiations.[212]

It is perhaps through the preparatory work for policy proposals that the secretariats have contributed (and could contribute) most to the integration of the two regimes. The documents released by the CBD secretariat in early 2012 underline this approach: they contain 'suggested recommendations' for the CBD's Subsidiary Body for Scientific, Technical and Technological Advice, which in turn contain suggestions for the CBD COP,[213] and eventually were largely incorporated as an annex in a COP decision.[214] Although such suggestions are common practice in the CBD process and the text as proposed need not be adopted by the CBD's treaty

---

207 CBD Decision X/33, para. 13(a).

208 CBD 2012a, 12–13.

209 Notably the (non-legally binding) Aichi targets, part of the Strategic Plan for Biodiversity 2011–2020, as set out in CBD Decision X/2, Annex.

210 CBD 2012a, 13–25.

211 CBD 2012a, 13.

212 CBD 2012c.

213 CBD 2012a; see also CBD 2012b.

214 CBD Decision IX/16, Annex.

bodies, it is notable that by providing draft texts which contain important ideas for further integration, the CBD secretariat exerts cognitive influence on the decision-making process on regime overlaps.

5.2 Enhancing Institutional Coordination

Although institutional coordination to address the interactions between the climate and biodiversity regimes is clearly intensifying (even if it still mainly takes place at the insistence of the CBD bodies), it remains to be seen whether enhancing institutional coordination is possible. A variety of options is conceivable. Some of these options are inspired by progress in other areas of international environmental law and governance, such as the clustering of the three chemicals conventions (see Chapter 5).

Several possibilities for enhanced institutional coordination have already been raised in the meetings of the decision-making bodies or the subsidiary bodies of the conventions and have also been carried out in practice. These options include collaboration among national focal points; collaboration among the subsidiary bodies of the Convention; the development of joint work programmes or other medium- to long-term planning documents; joint workshops; joint capacity-building initiatives; joint publications; guidance for joint scientific assessments; exchange of information and experiences; and cooperation on communication, education and public awareness.[215] It might be possible to formalize these activities by concluding a Memorandum of Understanding/Cooperation between the secretariats of both regimes, which would fall within the mandate of the secretariats. Coordination could be further strengthened by harmonizing reporting requirements or the timing of reporting, or by streamlining guidance to the financial mechanisms, in particular the Global Environment Facility.[216]

A more far-reaching option would be to hold a simultaneous meeting of the decision-making bodies of the two conventions, possibly joined by the UN Convention to Combat Desertification.[217] While this option has been raised, a joint meeting at the United Nations Conference on Sustainable Development in 2012 (Rio+20) was considered unfeasible given the 'procedural, practical and operational difficulties and the

215 CBD 2004, para. 9; see also CBD 2012b, 5–6.
216 CBD 2004.
217 von Moltke 2005, 179–180; Fauchald 2010, 16.

political risks involved'.[218] Still, it is notable that an informal joint meeting of the treaties' subsidiary bodies has been possible in the past.[219]

Yet another proposal is to cluster MEAs, an option that comprises some of the above-mentioned proposals,[220] and that has been pursued in the case of the chemicals conventions. Konrad von Moltke describes clustering as '[i]nstitutional and organisational arrangements short of a merger that will increase the efficiency and effectiveness of existing agreements without requiring elaborate changes in legal or administrative arrangements'.[221] Clustering could entail the grouping of MEAs by issue area (e.g., an atmosphere or biodiversity cluster), by region (e.g., a Europe or South-East Asia cluster), by function (e.g., implementation review, compliance or the transfer of finance and technology), by human activity (e.g., transport or industrial production) or by environmental policy instrument (e.g., trade restrictions).[222] In the climate–biodiversity case, the most sensible clusters would relate to common functions carried out by the climate and biodiversity regimes, but they could possibly also be related to the activities governed or the policy instruments used. However, there are important legal, practical and political impediments to any such proposal. Any cluster involving the climate treaties would require specific endorsement by the UNFCCC COP and the Kyoto Protocol COP/MOP (which would in turn require consensus), creating a significant legal hurdle. Moreover, from a practical perspective, the large size of the climate regime (in terms of its budget and rulebook) makes it likely that it would dominate any cluster in which it was involved.[223] From an administrative perspective, clustering could involve merging the secretariats of the CBD and the UNFCCC in a way similar to the chemicals conventions.[224] This raises obvious practical and political difficulties in the climate–biodiversity case, with the CBD secretariat being located in Montreal (Canada) and the UNFCCC secretariat in Bonn (Germany) and with the former linked to UNEP and the latter to the UN

---

[218] Comments by UNFCCC Executive Secretary Christiana Figueres, in Joint Liaison Group 2011, 5.

[219] http://www.cbd.int/programmes/cross-cutting/climate/joint.shtml (accessed 11 December 2013).

[220] UNEP 2001, para. 26; see also Oberthür 2002; von Moltke 2005.

[221] von Moltke 2005, 177–178.

[222] Oberthür 2002, 328–329; Biermann 2005, 137–138.

[223] Fauchald 2010, 35.

[224] The chemicals conventions appointed a joint head of the secretariats and identified joint services to be carried out by the secretariats. See K. Scott 2011, 207.

more broadly. Moreover, common housing was already attempted in the 1990s, at the initiative of UNEP, but failed to produce the expected proximity-related benefits.[225]

Another set of options for enhancing institutional coordination relates to the strengthening of the coordinating role of UNEP with respect to existing MEAs, for instance by making it a specialized agency rather than a programme in the UN system.[226] UNEP's role is seen as important as its mandate originally included the coordination of environmental activities in the UN system.[227] At the Rio+20 summit, agreement was reached to strengthen UNEP, including its coordination role in the UN system.[228] However, a key problem in the climate–biodiversity case is that the UNFCCC is not formally linked to UNEP. Without an agreed formal relationship, it is difficult to see how UNEP could coordinate the relationship between the climate and biodiversity regimes. Moreover, '[i]nstitutionally, the convening power of UNEP is dwarfed compared to other institutions dealing with major environmental concerns such as climate change',[229] implying that any coordination between the climate regime and other environmental regimes is best carried out by actors in the climate regime itself.[230]

For Cinnamon Carlarne, addressing the climate–biodiversity overlap would require an international environmental organization in the long run.[231] Her proposal relates to an ongoing debate about whether a World Environment Organization or UN Environment Organization is needed.[232] Proponents of a World Environment Organization argue that it could improve coordination among MEAs, facilitate their implementation at the national level and provide incentives for financial and technology transfer to developing countries.[233] However, creating a World Environment Organization would likely provoke resistance from existing international environmental regimes, including the climate and biodiversity regimes, and the parties providing financial support might be less willing to

---

225 Hicks 1999, 1665.

226 It is beyond the scope of this book to conduct an assessment of the various options for the institutional reform of international environmental governance. See, e.g., Fauchald 2010; Bernstein and Brunnée 2011.

227 Ivanova 2012, 572.

228 UNGA 2012, para. 88(c).

229 Inomata 2008, 30.

230 See also Fauchald 2010, 42–43.

231 Carlarne 2008, 473.

232 Biermann and Bauer 2005.

233 Biermann 2005, 120–127.

transfer control over the funding mechanisms.[234] In addition, the coordinating functions carried out by such an organization could arguably also be simply carried out by the individual institutions concerned. In other words, joint interaction management could in theory deliver the same results as collective interaction management.

More generally, there are clear barriers to enhancing institutional coordination between the climate and biodiversity regimes. An important one is that the US is a party to the UNFCCC, but not to the CBD (or to the Kyoto Protocol). A broad mandate for the climate regime's treaty or administrative bodies to coordinate with the CBD could lead to the perception that state sovereignty is eroded by 'importing' concepts or rules from the CBD.[235] Party submissions to the UNFCCC confirm this fear. In the process of creating the Joint Liaison Group, China noted that the group should not lead to 'dilution or modification of the obligations, interests or rights of the Parties under conventions'.[236] Commenting on the paper on options for enhanced cooperation prepared by the Joint Liaison Group in 2004, the US noted that the different Rio Conventions 'have a distinct legal character, mandate and membership'.[237] Going one step further, Australia argued that '[t]he CBD and the [UN Convention to Combat Desertification] do not have a legitimate role in greenhouse [gas] mitigation, which is clearly the work of the UNFCCC'.[238] Australia's response shows that the matter of sovereignty loss is related to the potential consequences of further integration of the two regimes: parties to the climate regime may not be ready to give biodiversity conservation a more prominent place at the expense of achieving rapid and cost-effective emission reductions. The cautious responses could also point to a fear among parties in the climate regime of overburdening the climate negotiations with considerations that are not directly related to emission reductions. This fear need not materialize, however, as it might be possible to introduce safeguards in stages rather than all at once.[239]

From the perspective of the biodiversity regime, the issue of incongruent memberships may also lead to concerns about institutional coordination, as parties to the CBD may feel that non-parties like the US may interfere in their business.[240]

---

234 von Moltke 2005, 177.
235 Wolfrum and Matz 2003, 163.
236 UNFCCC 2002, 7.
237 UNFCCC 2006b, 16.
238 UNFCCC 2006b, 5.
239 Gardner *et al.* 2012.
240 Fauchald 2010, 15.

Another key barrier to enhancing institutional coordination is that the treaty and administrative bodies of the climate and biodiversity regimes do not have clear legal authority to develop rules on overlapping issues. Moreover, the mandates of the cooperating bodies differ in their scope. For instance, at its fifth meeting, the Joint Liaison Group argued for consistent guidance from the various COPs, indicating that it can only facilitate, but not guarantee, this consistency.[241] Furthermore, at its ninth meeting, the Group lamented the fact that 'there remains a disconnect between the roles and mandates given to the [Joint Liaison Group] by each convention with this disconnect resulting in limitations when considering the implementation of the requested activities'.[242] Because of these limitations, the Group acts primarily as a forum to facilitate information exchange, and to encourage harmonizing implementation of the Rio Conventions at the national level.[243]

6. AUTONOMOUS INTERACTION MANAGEMENT

The legal techniques and institutional efforts discussed above are by no means without effect, but they are unlikely to be a panacea for managing interactions between the climate and biodiversity regimes. This section therefore examines the potential for autonomous interaction management by states and non-state actors.

6.1 States

In the implementation phase of two international environmental regimes, states have significant capacity to enhance synergies and mitigate conflicts. While the rules developed under the climate treaties may provide incentives that result in harm to biodiversity, the rules are broad and flexible enough to allow states to adopt policies and measures that seek to prevent such harm. Various policy instruments could specifically address the biodiversity impacts of climate change mitigation measures, although domestic circumstances and politics may influence the feasibility of such measures.[244]

---

[241] UNFCCC 2004, para. 4(l).

[242] Joint Liaison Group 2009, para. 11.

[243] Chambers 2008, 69.

[244] For instance, measures restricting the use of low-cost forestry projects that do not contribute to biodiversity protection may lower the overall cost-effectiveness of a country's climate policy. Whether a government is willing to

The domestic policymaker's toolkit is substantial.[245] For example, in terms of regulation, states could adopt mandatory biodiversity standards for forestry activities within their jurisdiction.[246] Furthermore, even though the requirements to conduct environmental impact assessments at the international level are weak, states can adopt stringent impact assessment procedures or promote the use of strategic environmental assessments.[247] Whereas environmental impact assessments are normally used at the project level, strategic environmental assessments are aimed at assessing the environmental impacts of policies and measures.[248] Such assessments could therefore lead to more structural integration of biodiversity concerns in mitigation policies at the national level.[249] Policymakers could also complement their regulatory and procedural policies with informational measures, for example, by creating awareness among forest managers about the biodiversity impacts of mitigation activities or by introducing a mandatory certification scheme.[250]

Additionally, states could design their domestic emissions trading systems in a way that accounts for the biodiversity impacts of forestry projects. With regard to credits from CDM afforestation and reforestation projects (and perhaps also REDD credits in the future), states buying credits could decide whether they want to use such credits in their trading systems and, if so, to what extent. For example, the EU decided to fully exclude credits from LULUCF from its emissions trading system.[251] Likewise, the system does not include any provisions on the use of REDD credits. While no emissions trading scheme has been adopted in the US, proposed bills have included detailed provisions on the use of REDD offsets in a domestic trading scheme. For instance, the Waxman-Markey bill – which was passed by the House of Representatives, but never made it through the Senate – included provisions requiring the US

make this trade-off will depend on many factors, including the country's economic situation, other available mitigation options and the strength of environmental and industrial lobbies.

245 Sagemüller 2006, 232–240; Pistorius *et al.* 2010, 24–26.

246 Ebeling and Fehse 2009, 35–37.

247 Sagemüller 2006, 234.

248 Secretariat of the CBD 2003, 90.

249 Secretariat of the CBD 2003, 94.

250 Sagemüller 2006, 238.

251 Directive 2003/87/EC of the European Parliament and of the Council of 13 October 2003 establishing a scheme for greenhouse gas emission allowance trading, as amended by Directive 2004/101/EC, OJ [2004] L338/23, Art. 11a.3(b).

Environmental Protection Agency to ensure that REDD offsets fulfil certain minimum standards.[252]

States wishing to fund forestry activities through bilateral or regional initiatives could tie their funding to the achievement of certain biodiversity benefits, or the inclusion of biodiversity safeguards. Projects seeking funding from the World Bank's Forest Carbon Partnership Facility, for instance, need to undertake a strategic environmental and social assessment to be eligible for funding.[253] Furthermore, biodiversity co-benefits are one of the safeguards included in Norway's bilateral agreement with countries such as Indonesia. Although such conditionalities may be controversial and may be difficult to impose as 'hard' requirements, they could lead to investments that result in multiple benefits.

These are but a few examples of autonomous actions that states could undertake to ensure that they live up to their commitments under both the climate and biodiversity treaties, underlining the potential of state-driven autonomous interaction management.

6.2 Non-state Actors

While government action could provide incentives for domestic actors to change their behaviour in order to contribute to the objectives of the climate and biodiversity regimes, those incentives may also be offered by private actors.

One of the key developments in this regard has been the growing number of voluntary standards for forestry projects.[254] In stark contrast to the regulatory market, forestry projects have a large share in the voluntary carbon market.[255] The prominence of forests in the voluntary carbon market has been accompanied by the emergence of several standards that address the biodiversity impacts of forest mitigation projects.

A notable example is the Climate Community and Biodiversity Alliance, a partnership of non-governmental organizations, businesses and research institutions, which has developed standards for terrestrial carbon mitigation projects, addressing the various safeguards of the UNFCCC.[256] Other examples of private standards that address the non-carbon effects

---

[252] American Clean Energy and Security Act of 2009, H.R. 2454, 111th Cong. (2009), Sec. 754(d)(5).

[253] Ebeling and Fehse 2009, 36; Pistorius *et al.* 2010, 10.

[254] Hamilton *et al.* 2008, 297–305.

[255] Hamilton *et al.* 2008, 292.

[256] E.g., Visseren-Hamakers *et al.* 2011, 98–99.

of REDD activities include the Global Conservation Standard, the CarbonFix standard and the Plan Vivo standards.[257] Together with CARE International, the Climate Community and Biodiversity Alliance is also involved in the REDD+ Social and Environmental Standards initiative. Unlike the Alliance's standards, standard development under this initiative also involves several public actors at the national and sub-national levels.[258]

Voluntary standards have been increasingly linked to the regulatory market. The CDM Gold Standard, for example, started as an initiative by a non-governmental organization, but has gradually become an important instrument for promoting sustainable development in the CDM.[259] Forestry projects are, however, so far not eligible to receive Gold Standard certification.

While the scale of the voluntary market, compared to the current and the potential regulatory market, is relatively small, it is notable that for reputational or other reasons, the demand for biodiversity benefits in forest carbon projects is significant.[260] Moreover, there seems to be a willingness on the part of buyers to pay a higher price for credits resulting from projects that have been certified.[261] In other words, there are incentives to integrate biodiversity and climate concerns in the voluntary market. The drawback is that the voluntary market will not deliver these biodiversity benefits on a large scale.[262]

These examples show how autonomous interaction management can also be driven by non-state actors. The activities by non-state actors could provide valuable experience that could help determine how two regimes might work together in practice. Ideally, such practical experiences would be integrated with intergovernmental efforts to manage interactions. For example, Andrew Long discusses how non-state certification initiatives could work hand-in-hand with government efforts to ensure that REDD simultaneously contributes to climate, biodiversity and human welfare objectives.[263] In the absence of structural solutions at the global level, autonomous management could thus lay the groundwork for mutually supportive environmental treaties.

---

257 Merger *et al.* 2011.
258 http://www.redd-standards.org/ (accessed 11 December 2013).
259 Levin *et al.* 2009.
260 Karousakis 2009, 20.
261 Ecosecurities 2010, 28.
262 Ebeling and Fehse 2009, 49.
263 Long 2011; see also Levin *et al.* 2009, 791.

7. CONCLUSIONS

This chapter has examined the consequences and management of interactions between two international environmental regimes, focusing on interactions between the climate and biodiversity regimes. In the absence of a comprehensive international legally binding instrument on forests, both regimes have sought to fill a niche in global forest governance, but they have done so from different perspectives. Whereas the biodiversity regime has adopted an ecosystem-based approach, the climate regime has viewed forests primarily as sinks or sources of emissions. The climate regime has dominated rule development on aspects that concern both regimes through decisions on forest carbon. The resulting rule complex triggered behavioural interactions that led to a policy conflict in the implementation phase, but there is clearly also a large potential for synergies, especially following the emergence of REDD in the climate negotiations.

With respect to legal techniques, the chapter shows that there is potential for treaty changes, especially given the window of opportunity provided by the ongoing climate negotiations, but also argues that it is more likely that changes to accommodate biodiversity concerns will be implemented by means of a COP decision. In addition, the chapter highlights the positive role treaty interpretation could play. The discussion of legal techniques also underlines that there are inherent limitations to the use of conflict prevention and resolution rules in managing interactions in international environmental law. These limitations are primarily due to the fact that such rules disregard the interactions triggered by treaty body decisions. Legal techniques are hence limited by the fact that international environmental regimes are often based on 'living treaties'. Moreover, legal techniques also disregard the potential for synergies between international environmental regimes.

Institutional coordination could complement legal approaches in addressing interactions between international environmental regimes. In the climate and biodiversity context, these efforts have mainly been undertaken by actors in the biodiversity regime, including the COP and its secretariat. While these initiatives have been important for creating awareness of climate–biodiversity interlinkages, synthesizing research on this issue and generally fostering cooperation, they have not yet managed to address the tensions about the use of sinks in mitigation activities.

Many options for enhanced coordination between international environmental regimes are possible beyond these first steps, ranging from options to increase efficiency, such as harmonized reporting and joint guidance, to financial mechanisms, to more radical changes, such as clustering or the creation of overarching institutional arrangements, such as a World Environment Organization. In addition to legal and practical difficulties associated with these options, there are more general limitations that complicate enhanced coordination of the biodiversity and climate regimes. These include the lack of parallel memberships of the regimes and the unwillingness of parties in the climate regime to allow parties in the biodiversity regime to influence rule development. Institutional coordination has also been constrained because states have been reluctant to cede coordination authority to treaty bureaucracies, as was exemplified by the Joint Liaison Group.

In the absence of effective collective interaction management, it is still possible to mitigate conflicts and enhance synergies through autonomous interaction management by state and non-state actors. The chapter has identified various existing and proposed initiatives, all of which have the potential to tackle the problems of climate change and biodiversity in conjunction. Only time will tell whether such efforts can realize this potential and, if so, whether and how they could be scaled up to the international level.

8. The UN climate regime and the World Trade Organization

1. INTRODUCTION

The last case study of this book examines the interactions between the UN climate regime and the World Trade Organization as a case of interactions between an international environmental regime and an international economic regime. Specifically, the chapter seeks to analyse the consequences of interactions between the climate and trade regimes and to examine different ways of managing the relationship between the two regimes. The chapter also aims to provide insights into options for dealing with one of the most politically sensitive issues at the intersection of the climate and trade regimes: unilateral climate-related trade measures.

The close connection between greenhouse gas emissions and a wide array of economic activities necessitates the consideration of linkages between the climate and trade regimes. Despite the fact that many observers have discussed a possible clash between the two regimes, the multilateral trading system and the climate regime seem mutually supportive at first glance. However, this relationship is likely to change in the future due to the growing acknowledgement that abating climate change requires deeper emission cuts. More specifically, the use of unilateral climate-related trade measures by participants in the UN climate regime may lead to conflicts with the principles and rules of the WTO.

2. THE GOVERNANCE OF CLIMATE CHANGE AND TRADE

2.1 Climate Change and Trade

Current and projected human-induced climate variability is linked to economic patterns, which are responsible for the majority of greenhouse gas emissions. If climate change is to be effectively addressed, solutions will necessarily affect the global economy.

International trade affects climate change in at least three different ways.[1] First, it does so by potentially increasing economic activities that may in turn lead to rising emissions. In addition to this 'scale' effect, trade liberalization may also change emissions through a 'composition' effect. This refers to a change in emissions as a result of sectoral specialization in an area in which a country has a comparative advantage. If such an area includes emissions-intensive industries (e.g., steel or cement), emissions are likely to rise. Conversely, if a country has a comparative advantage in cleaner industries, its emissions presumably will go down. The third effect that can be distinguished is the 'technique' effect, which concerns the changes in emissions in the production process. This includes reduced emissions resulting from the increased trade in climate-friendly goods and services, as well as the diffusion of technologies by means of which these are produced. The extent to which these combined effects have resulted in changes in emissions in practice is not entirely clear. While various studies show that the scale effect dominates the other effects, indicating that trade leads to an overall increase in emissions, other studies have pointed out that the outcomes may lead to emission reductions for developed countries yet emission increases for developing countries.[2]

Trade may also have a direct impact on greenhouse gas emissions as goods are transported from one country to another. Different means of transportation result in different levels of emissions. For instance, international air transport is known to lead to more emissions than maritime transport. While the latter is still the main mode of international transportation, the share of air transport is increasing rapidly, meaning that direct emissions from trade are also likely to rise.[3]

While trade may thus affect climate change, climate change may also have an effect on international trade. Climate impacts may alter trade flows by changing countries' comparative advantages. For example, countries exporting agricultural products may be affected by increasing droughts, while countries depending on tourism may also see a change in their income.[4] Furthermore, the trade infrastructure itself may be affected by climatic changes. The *Stern Review* in particular noted the risks for

---

1 UNEP and WTO 2009, 49–52.
2 UNEP and WTO 2009, 53–54, 61–62.
3 UNEP and WTO 2009, 58–59.
4 UNEP and WTO 2009, 64.

sea-borne trade resulting from rising sea levels and rapidly increasing temperatures in the Polar regions.[5]

Finally, the relationship between climate change *policies* and trade has been the subject of extensive discussions, as I will show in this chapter. In particular, policies and measures taken to achieve climate policy goals may directly or indirectly restrict trade flows between countries. Furthermore, specific measures to reduce emissions might adversely affect competitiveness and hence reduce countries' willingness to participate in such measures.[6]

2.2 The World Trade Organization and Climate Change

The WTO in brief

The WTO is the only global organization that deals with the rules of trade between nations. It formally came into being on 1 January 1995 as a result of the Uruguay Round of trade negotiations. The WTO agreements, signed by the large majority of trading nations, are at the heart of the organization. The WTO not only intends to further the implementation, administration and operation of its agreements, but also serves as a forum for further negotiations on the liberalization of trade in goods and services. In addition, the WTO seeks to resolve disputes between its members through its relatively strong dispute settlement mechanism.[7] The preamble to the Agreement Establishing the WTO contextualizes these objectives, adding that trade relations should lead to

> raising standards of living, ensuring full employment and a large and steadily growing volume of real income and effective demand, and expanding the production of and trade in goods and services, while allowing for the optimal use of the world's resources in accordance with the objective of sustainable development, seeking both to protect and preserve the environment and to enhance the means for doing so in a manner consistent with their respective needs and concerns at different levels of economic development.[8]

Against this background, various agreements aim at lowering different barriers to trade and market access. The most important agreement is the General Agreement on Tariffs and Trade (GATT), which is concerned with the international trade in goods. The two other main agreements are the General Agreement on Trade in Services (GATS) and the TRIPS

---

5 Stern 2006, chapter 5, 17.

6 Charnovitz 2003, 141.

7 Agreement Establishing the WTO, Art. III.

8 Agreement Establishing the WTO, preamble, recital 1.

Agreement, dealing with the international trade in services and intellectual property rights, respectively. Other WTO agreements concerned with reducing barriers to the trade in goods include the Agreement on Technical Barriers to Trade (TBT Agreement), which deals with the international harmonization of technical regulations and standards, and the Agreement on Sanitary and Phytosanitary Measures, which seeks to ensure that health and safety measures are not being used to protect domestic producers.

A number of interrelated fundamental principles guide the multilateral trading system, including the principles of predictability and stability, transparency, the promotion of fair competition and, above all, non-discrimination. The latter principle is expressed in two well-known rules. First, a country should not discriminate between producers from other member countries and domestic producers – the 'national treatment' rule.[9] This means that products that are imported and domestic products that are 'like' should be treated as such. The second rule is that a country should not discriminate between its trading partners – the 'most-favoured nation' rule.[10] The purpose of this rule is to ensure that like products are treated alike, irrespective of their origins or destination.

In terms of organizational structure, the highest WTO authority is the Ministerial Conference, which meets at least every two years.[11] It can adopt measures on all matters under any of the WTO agreements. In the intervals between the meetings of the Ministerial Conference these functions are carried out by the General Council.[12] The General Council also acts as the Dispute Settlement Body and, in that guise, may establish a Panel for disputes between WTO members.[13] The third main organ is the WTO secretariat, headed by the Director General, which offers administrative and technical support.[14]

The WTO and the environment

The WTO's attitude towards environmental issues has been the subject of discussions since its inception. A short recapitulation of the 'trade and

---

[9] GATT, Art. III.
[10] GATT, Art. I.
[11] Agreement Establishing the WTO, Art. IV.1.
[12] Agreement Establishing the WTO, Art. IV.2.
[13] Agreement Establishing the WTO, Art. IV.3.
[14] Agreement Establishing the WTO, Art. VI.

environment' debate provides an indication of how the WTO has approached the issue of climate change.[15]

Although measures to protect the environment may limit the free flow of traded goods and services, they can be justified under several WTO agreements.[16] A distinction needs to be made between unilateral measures adopted by a WTO member, and multilateral measures adopted under an MEA. While there is an emerging body of case law under the WTO's dispute settlement mechanism on the former, to date no multilaterally agreed trade-related environmental measure has been challenged, let alone ruled out, under WTO law.

Most attention on the legal overlap between the WTO and environmental policies has focused on unilateral measures aimed at extraterritorial protection of the environment. This discussion has been framed by a series of high-profile cases before the WTO's Dispute Settlement Body (and its predecessor under the GATT) in which the legality of trade-restrictive measures with an environmental rationale was challenged. Various cases have sought to clarify the scope of the GATT's limited environmental exceptions, which can justify trade-restrictive measures 'necessary to protect human, animal or plant life or health';[17] or 'relating to the conservation of exhaustible natural resources if such measures are made effective in conjunction with restrictions on domestic production or consumption',[18] provided that 'such measures are not applied in a manner which would constitute a means of arbitrary or unjustifiable discrimination between countries where the same conditions prevail, or a disguised restriction on international trade'.[19]

The first trade and environment cases in the early 1990s (related to an import ban imposed by the United States on tuna caught in ways harmful to dolphins) resulted in rulings against extraterritorial environmental protection by the Panel under the GATT.[20] Following the establishment of the WTO, the dispute settlement system arguably shifted increasingly towards accommodating unilaterally adopted environmental measures. Commentators generally cite the seemingly more environmentally friendly outcomes of the shrimp-turtle disputes, in which the Appellate

---

[15] E.g., Esty 1994; Schoenbaum 1997; Charnovitz 2007; Vranes 2009; Gehring 2011.

[16] E.g., GATT, Art. XX; TBT Agreement, Art. 2.2; TRIPS Agreement, Art. 27.2.

[17] GATT, Art. XX(b).

[18] GATT, Art. XX(g).

[19] GATT, Art. XX, chapeau.

[20] Note that these reports were not adopted by the WTO members.

Body in two separate rulings outlined how unilateral trade measures imposed by the US could be allowed under certain circumstances.[21] In the first ruling, the Appellate Body invoked the concept of sustainable development in its determination of the scope of the environmental exceptions of Article XX of the GATT. In the second ruling, the Appellate Body further clarified the importance of entering into serious negotiations before adopting unilateral measures, emphasizing that such negotiations need not lead to an agreement provided they are conducted seriously and in good faith.[22] In subsequent disputes, the Appellate Body refined its approach to Article XX of the GATT, introducing various steps for evaluating whether environmental measures can be justified by it.[23]

Not all the discussions about the WTO's effects on environmental policies are about determining the legality of such policies. Political scientists have noted that the WTO exerts influence on environmental policymaking through a 'chilling' effect.[24] This means that the mere existence of the international trade regime (and the potential for conflict with it) might deter negotiators from adopting trade measures under environmental treaties. Although negotiators may refrain from doing so for many reasons, for example because they may entail an undesired 'hardening' of environmental treaties or reduce incentives for participating in a treaty,[25] there is nonetheless some empirical evidence of the chilling effect.[26]

Finally, the relationship of the WTO to the environment is also reflected in organizational developments.[27] The WTO has addressed environmental issues, initially through adopting a Ministerial Decision on Trade and Environment in 1994, and the establishment of the Committee on Trade and Environment (CTE) in 1995.[28] The CTE is open to all WTO members, while a number of observers from intergovernmental organizations are also represented. While the CTE provides the main

---

[21] *United States – Import Prohibition of Certain Shrimp and Shrimp Products*; *United States – Import Prohibition of Certain Shrimp and Shrimp Products, Recourse to Art. 21.5 DSU*.

[22] *United States – Import Prohibition of Certain Shrimp and Shrimp Products, Recourse to Art. 21.5 DSU*, para. 176.

[23] *European Communities – Measures Affecting Asbestos and Asbestos-Containing Products*; *Brazil – Measures Affecting Imports of Retreaded Tyres.*

[24] Eckersley 2004.

[25] Gehring 2011, 236.

[26] Axelrod 2011.

[27] Charnovitz 2007, 690–692.

[28] WTO 1994.

WTO forum for the discussion of trade–environment linkages, its usefulness has been questioned.[29] The relationship between trade and environment further received notable attention in the Doha Ministerial Declaration of November 2001,[30] which launched the current round of trade negotiations. The Declaration calls for negotiations on various issues, such as the relationship between MEAs and trade, eco-labelling and the effects of environmental measures on market access, especially for developing countries. These negotiations are taking place in Special Sessions of the CTE. However, the prospects for agreement on the relationship between trade and environmental policies are gloomy, mainly because the entire Doha Round came to an almost complete halt in 2008.[31]

The WTO and climate change

The CTE has not engaged with interactions between climate change and international trade in depth.[32] Since 2007, however, the WTO's attitude towards climate change has appeared to be changing. More specifically, several members are increasingly willing to discuss trade–climate overlaps within the CTE.

In the context of the Doha Round, the most eye-catching developments have been in the CTE Special Sessions negotiations on liberalization of environmental goods and services.[33] Several WTO members have drafted proposals to liberalize trade in specific climate-friendly goods and services, including notably a joint proposal tabled in 2007 by the EU and the US.[34] The proposal, which suggested introducing zero tariffs for 43 climate-friendly products, was heavily criticized by developing countries, who pointed out that the list was biased against their exports.[35]

In addition, the CTE started to address issues within its mandate that related to trade and climate change in 2008, following an Informal Trade Ministers Dialogue on Climate Change Issues organized in Bali the year

---

[29] Charnovitz 2007, 690.

[30] WTO 2001, paras. 31–33.

[31] The Doha Round was partly reignited at the trade ministerial in Bali in December 2013. This may lead to new discussions on environmental goods and services. See *The Economist* 2013.

[32] One of the tangible outputs is the regularly updated matrix of multilateral environmental agreements. See, e.g., WTO 2011a, 107–108.

[33] WTO 2001, para. 31(iii).

[34] http://trade.ec.europa.eu/doclib/docs/2007/november/tradoc_136955.pdf (accessed 11 December 2013).

[35] International Centre for Trade and Sustainable Development 2007.

before.[36] The dialogue brought together ministers and trade officials from over 30 countries.[37] It generally underlined the potential for mutually supportive regimes, and called for continued high-level engagement on the issue. However, the ministers also acknowledged that there is potential for conflict, especially in the absence of a multilateral climate change agreement.

The potential for 'win-win' solutions was emphasized by former WTO Director General Pascal Lamy, who argued that the WTO could support international climate policy by making available some of its tools, for example by reducing fossil fuel subsidies. Although identifying a potential role for the WTO, Lamy also stressed that combating climate change requires agreement under the UNFCCC, thereby showing wariness of unilateral trade measures.[38] The 'win-win' framing of climate and trade policies has played an instrumental role in the increased activity by the WTO secretariat on trade and climate change. Although it is difficult to discern the precise motives for Lamy's framing, the rationale might be 'the idea that the WTO will benefit if [it] appears greener to the public and that climate negotiations will benefit from a signal that WTO rules are not an excuse for inaction on climate'.[39]

2.3 The Climate Regime and Trade

In contrast with the WTO agreements, the climate treaties contain an explicit statement about their relationship with the trade regime. In what could be deemed an example of the chilling effect,[40] the UNFCCC appears to defer to the trade regime, using language lifted straight from the GATT. Neither the UNFCCC nor the Kyoto Protocol contains trade measures, unlike, for example, the Montreal Protocol, which explicitly calls for a trade ban with non-parties.[41] Instead, the climate treaties call for coherence of climate policies with international trade law. The UNFCCC states that '[m]easures taken to combat climate change, including unilateral ones, should not constitute a means of arbitrary or unjustifiable discrimination or a disguised restriction on international trade'.[42] According to Dan Bodansky, it thereby 'neither condones nor

[36] WTO 2009a, 13–19.
[37] WTO 2008, 1.
[38] Lamy 2007.
[39] Charnovitz 2010a, 275.
[40] Zelli and van Asselt 2010, 88; Gehring 2011, 236.
[41] Montreal Protocol, Art. 4.
[42] UNFCCC, Art. 3.5.

forbids using trade measures'.[43] The Kyoto Protocol reaffirms the commitment to 'minimize adverse effects on ... international trade' in the pursuit of its objectives.[44]

Robyn Eckersley points out that the WTO and the climate treaties actually share four important features: (1) the endorsement of sustainable development; (2) accounting for the special circumstances of developing countries; (3) support for an open economic system; and (4) supporting continued economic growth.[45] She characterizes the approach of the climate regime towards the trade regime as a 'minimalist or default approach to trade-environment integration'.[46] In this way, open conflicts between the two regimes have been averted, and the idea of mutually supportive agreements has been promoted.

However, as I will show in this chapter, the harmonious relationship between the climate and trade regimes may come under increased pressure, particularly if unilateral trade measures are adopted. Indeed, the mere prospect of unilateral trade measures has led to heightened attention in the climate negotiations in recent years.

3. CONSEQUENCES OF INTERACTIONS

This section first provides a (non-exhaustive) overview of the consequences of key interactions between the climate and trade regimes, focusing on: (1) the Kyoto Protocol's flexibility mechanisms; (2) domestic trade-related policies and measures; (3) the transfer of climate-friendly goods and technologies; and (4) preferential trade measures. It shows that there are several potential conflicts between the regimes but argues that at the same time there is room for synergies. The second part of this section shows that the prospect of unilateral trade measures, notably in the form of border adjustment measures (BAMs), may lead to a more conflictive relationship between the two regimes.[47]

---

43 Bodansky 1993, 505.
44 Kyoto Protocol, Art. 2.3.
45 Eckersley 2009, 12.
46 Eckersley 2009, 17.
47 This section draws on van Asselt and Brewer 2010; and Zelli and van Asselt 2010.

3.1 Consequences of Key Climate–Trade Interactions

Flexibility mechanisms

The Kyoto Protocol's flexibility mechanisms raise several trade issues, which so far have remained unresolved. First, because international emissions trading under the Protocol is limited to industrialized countries that have ratified the treaty, it 'implicitly prevents parties not included in Annex B from acquiring, issuing, or transferring emissions allowances under the Protocol'.[48] This restriction could be considered a form of trade discrimination, as it effectively excludes the large majority of developing countries, as well as non-parties to the Kyoto Protocol (i.e., the US).[49] These imbalances could amount to a violation of the WTO's non-discrimination rules if emission allowances could be defined as either 'goods' or 'products' under the GATT or as 'services' under the GATS.[50] However, some scholars have argued that allowances should be considered neither of those, and that the economic or financial value of allowances alone does not make them goods or services, similar to other entities such as electricity, oil or money, which also do not fall under GATT or GATS requirements.[51] Given this ambiguity, it is impossible to decide whether the case of trade restrictions in emission allowances constitutes an actual normative conflict between the climate and trade treaties.[52]

Even though there may be no such normative conflict, there are further implications of the use of flexibility mechanisms, which at the very least point to potential policy conflicts. Various design options may be chosen for a domestic emissions trading system, some of which have trade implications.[53] Depending on the design of the system, measures taken in the carbon market might affect the trade in goods and services in existing markets. Kyoto parties may restrict access to participation in international emissions trading through domestic measures, or may not recognize emission allowances generated in non-Kyoto parties.[54] For instance, the EU's emissions trading scheme at present does not recognize allowances from regional trading systems in the United States, although agreements

---

48 Werksman 2001, 170.
49 Voigt 2008b, 55.
50 Jinnah 2003, 721; Voigt 2008b, 55–56.
51 Werksman 2001.
52 Brewer 2003.
53 Werksman 1999b; van Asselt and Biermann 2007.
54 Voigt 2008b, 55.

to this end may be concluded in the future.[55] Furthermore, 'brokerage, consulting and insurance services associated with emissions trading could be considered commercial services within the normal meaning of the term',[56] and thus potentially fall under GATS rules. In fact, given the variety of services which can be involved in an emissions trading scheme, some scholars expect the GATS, rather than the GATT, to be applicable to emissions trading.[57]

Another design option that could be at odds with WTO law is the allocation of emission allowances under a domestic trading system. Certain allocation methods could be interpreted as the favourable treatment of a domestic industry over foreign competitors, in particular if a trading scheme stipulates the free-of-charge distribution of allowances, which could then be resold at a profit. Such a free allocation of financial assets might be classified as a subsidy.[58] In particular, the case of over-allocation (i.e., more allowances are distributed than needed) could constitute such an unfair subsidy. This situation is not hypothetical, as the widely reported over-allocation of emission allowances in the EU emissions trading system testifies.[59] The Agreement on Subsidies and Countervailing Measures (SCM Agreement) provides the main rules on subsidies. According to the treaty, subsidies need to entail a 'financial contribution by a government or any public body' or 'any form of income or price support' that confers a benefit.[60] A key question is whether measures are defined as 'prohibited' or 'actionable' subsidies under the agreement. Prohibited subsidies are contingent upon export performance or upon the use of domestic over imported goods.[61] For example, subsidies based on the use of domestic, low-carbon-emitting products over high-carbon-emitting imports are not allowed.[62] Actionable subsidies are subsidies that are specific (aimed at certain enterprises or industries[63]) and that cause injury to the domestic industry of another member, nullification or impairment of the benefits accrued by another

---

55 Directive 2003/87/EC of the European Parliament and of the Council of 13 October 2003 establishing a scheme for greenhouse gas emission allowance trading, as amended by Directive 2009/29/EC, OJ [2009] L140/63, Art. 25.1a and Art. 25.1b.

56 Brewer 2003, 337.

57 Petsonk 1999; Martin 2007.

58 Howse and Eliason 2009, 56–57.

59 van Asselt 2010, 136–137.

60 SCM Agreement, Art. 1.1.

61 SCM Agreement, Art. 3.

62 Assunção and Zhang 2002.

63 SCM Agreement, Art. 2.

member under the GATT, or serious prejudice to the interests of another member.[64] If a subsidy is not specific, it is non-actionable. Matthias Buck and Roda Verheyen have argued that the initial allocation does not amount to an actionable subsidy, since it can neither be regarded as a 'financial contribution' nor as 'income or price support' under the SCM Agreement, and since the emission allowances correspond to an obligation to only emit greenhouse gases under a certain cap. Unless this cap is set at a very generous level, the initial allocation would thus not likely be seen as a subsidy.[65]

Finally, the CDM may create tensions with international trade law, although given the relatively small scale of the potential conflicts,[66] this issue has received only scant attention to date. The main issues concern the potential incompatibility of CDM implementation within host countries with the WTO Agreement on Trade-Related Investment Measures (TRIMS Agreement). This could include the imposition of performance requirements mandating the use of local goods or services to ensure that CDM projects contribute to sustainable development.[67] While such a requirement would contribute to the CDM's objectives,[68] it may violate the TRIMS Agreement. However, the main potential interactions between the CDM and international investment law relate not to the TRIMS Agreement, but rather to (bilateral and regional) investment agreements concluded outside the auspices of the WTO.[69]

Another potential CDM-related trade barrier concerns possible differentiation between developing countries under the CDM. For instance, under the current Kyoto rules, certain least-developed countries receive preferential treatment because they have been largely bypassed by the CDM in the first years of its functioning.[70] Furthermore, some jurisdictions limit the access of countries like China and India to the mechanism, for instance by targeting technologies used in CDM projects in these countries.[71] While such proposals may discriminate between developing countries, it is questionable whether such discrimination

---

64 SCM Agreement, Art. 5.

65 Buck and Verheyen 2001.

66 Werksman 2009a, 323.

67 Werksman *et al.* 2003; Rechsteiner *et al.* 2009; Werksman 2009a.

68 Kyoto Protocol, Art. 12.2.

69 E.g., Werksman *et al.* 2003.

70 Kyoto Protocol Decision 2/CMP.5, paras. 23, 47–49.

71 See, for instance, Regulation 550/2011 on determining, pursuant to Directive 2003/87/EC of the European Parliament and of the Council, certain restrictions applicable to the use of international credits from projects involving

would be covered by the TRIMS Agreement, given that the agreement is aimed at the protection of goods connected to the investment rather than the investor itself.[72]

Trade-related policies and measures

The Kyoto Protocol commits Annex I Parties to adopting policies and measures through which they can reduce their emissions domestically. The treaty offers a (non-exhaustive) list of policies and measures that include, for instance, research, development and use of renewable energy and climate-friendly technologies; reduction or phasing out of fiscal incentives, tax and duty exemptions, and subsidies in greenhouse gas-emitting sectors; and limiting and reducing emissions in the transport sector.[73] However, the treaty does not specify the steps developed countries need to take to implement these policies and measures. As the policies and measures are not firmly defined, parties may implement them in ways that are trade distorting and/or violate WTO rules. There is a range of overlaps and potential clashes given the possible implementation of certain fiscal measures (subsidies, tariffs, taxes, or border taxes), regulatory measures (standards, technical regulations and labelling), and government procurement practices.

In line with the discussion above, some subsidies to firms for climate-friendly products, production processes, research, development or export might not be allowed under the SCM Agreement.[74] Such subsidies may impede market access of foreign countries and/or increase exports of the subsidizing country.[75] Although there may be some cause for concern that the WTO rules do not allow for the type of subsidies preferable from the climate change perspective,[76] '[i]f WTO Members ensure that their subsidies are widely available, also to foreign companies, it is likely that the subsidy will not be specific and will [not] be challengeable by other Members'.[77]

The other side of the subsidies coin is that a *reduction* of subsidies for fossil fuel producers or industrial emitters presents potential synergies

---

industrial gases, OJ [2011] L149/1. On differentiation in the CDM generally, see Bakker *et al.* 2011.

72 Rechsteiner *et al.* 2009, 310–311.

73 Kyoto Protocol, Art. 2.1.

74 Green 2006.

75 UNEP and WTO 2009, 115.

76 Green 2006.

77 Bernasconi-Osterwalder and Norpoth 2009, 29.

between international trade and climate policies.[78] Going one step further, Nobel laureate Joseph Stiglitz has suggested that the failure to internalize externalities presented by climate change should be seen as an unjustifiable subsidy.[79] Stiglitz had in mind that Kyoto parties would be able to take countervailing measures under the SCM Agreement. While some trade lawyers have been quick to dismiss this proposal,[80] others have not completely ruled out Stiglitz's suggestion.[81]

Governments might also choose to put burdens on energy-intensive foreign companies by adopting regulatory measures or by imposing taxes or other charges on their greenhouse gas-intensive imports. One major uncertainty about the WTO compatibility of such measures, particularly with respect to Article III of the GATT, relates to the question of processes and production methods (PPMs).

There is an extensive academic debate on whether products made using different PPMs are 'like' in the sense of the national treatment provision of the GATT.[82] Although the jury is still out on what exactly is permissible under WTO law, there is general consensus that discriminatory measures based on product-related PPMs are allowed,[83] whereas distinguishing on the basis of non-product-related PPMs is more controversial. Clear examples of measures based on product-related PPMs are emission standards for cars, the consequences of which are physically traceable in the final product and thus clearly matter from a consumer's perspective. An example of a measure based on a non-product-related PPM would be a regulation that requires that steel is produced using green electricity. While there may be valid reasons for prescribing such measures (social, environmental, moral, etc.), they do not leave a discernible trace on the final product.[84] As Steve Charnovitz notes, however, this 'related/non-related' distinction is not as clear-cut as it seems, as it is based on the (false) 'assumption that consumer preferences can be neatly divided between the physical characteristics of the product and a bundle of other ecological or moral concerns'.[85] Indeed, the question of 'likeness' under Article III does not depend on physical

---

78 Green 2006, 381; Bigdeli 2008.
79 Stiglitz 2006.
80 E.g., Bhagwati and Mavroidis 2007; Pauwelyn 2007, 14–16.
81 Howse and Eliason 2009, 74–76.
82 Howse and Regan 2000; Charnovitz 2002; Vranes 2009, 191–222 and 319–350.
83 UNEP and International Institute for Sustainable Development 2005, 54.
84 Cf. Charnovitz 2002, 65–66.
85 Charnovitz 2002, 66.

traceability, but rather on the prevailing competitive relationship.[86] In their judicial practice, the WTO dispute settlement bodies have consistently applied four criteria to identify the likeness of products: (1) the properties, nature and quality of the products; (2) the end uses of the products; (3) consumers' perceptions and behaviour in respect of the products; and (4) the tariff classification of the products.[87] This could mean that two products that are exactly the same, but which are made through processes resulting in different emissions, could still be considered 'unlike' depending on how consumers' tastes and habits are constructed. This determination could in turn be informed by the environmental impacts of PPMs, which, according to Robert Howse and Antonia Eliason, presents us with the greatest challenge 'of determining accurately whether a particular imported product is produced with significantly higher carbon emissions than a particular domestic product'.[88] Even if non-product-related PPMs are found to violate the non-discrimination rules of the WTO, this does not deem them automatically illegal, as they may be saved by recourse to the exceptions provided for in the GATT.[89]

Another type of domestic policy that might violate the non-discrimination principles of the GATT is the imposition of taxes at the border on energy-intensive imports (or the inclusion of importers in a domestic emissions trading system) from countries that are not party to the Kyoto Protocol or that do not take 'comparable' climate action. There is a long-standing legal debate on such border adjustment measures, with some experts holding that under certain circumstances they could be defended and sustained under WTO law,[90] and others warning against their protectionist implications and possible violation of the GATT.[91] Much will depend on the design of such measures, as I will discuss in greater detail below.

A different category of policies comprises product standards, labels and technical regulations establishing minimum requirements for goods on the basis of their energy or greenhouse gas intensity during production or use, which might conflict with the national treatment principle under

86 Vranes 2009, 324.

87 E.g., *European Communities – Measures Affecting Asbestos and Asbestos-Containing Products*, para. 101.

88 Howse and Eliason 2009, 68.

89 Charnovitz 2002, 63.

90 Biermann and Brohm 2005; de Cendra de Larragán 2006; Ismer and Neuhoff 2007; Pauwelyn 2007; Sindico 2009.

91 Quick 2008; Zane 2011.

the GATT or with the TBT Agreement.[92] The climate regime also permits certain government purchasing policies that might create tensions with the WTO Agreement on Government Procurement.[93] Yet altogether, labelling and standards (at least voluntary ones), and government procurement practices are unlikely to result in a conflict with WTO rules.[94]

Transfer of climate-friendly goods and services

A key example of potential synergy between the climate and trade regimes concerns the removal of trade barriers in favour of climate-friendly goods or services, and the development and transfer of low-emission technologies. The UNFCCC states that 'developed country Parties … shall take all practicable steps to promote, facilitate and finance, as appropriate, the transfer of, or access to, environmentally sound technologies and know-how to other Parties, particularly developing country Parties, to enable them to implement the provisions of the Convention'.[95] This provision is based on a broad understanding of technology transfer, which includes capacity building in the receiving countries (see Chapter 6). However, instead of facilitating knowledge transfer and capacity building, companies in developed countries have much higher incentives to build new technologies completely 'in house' to secure maximum profits and reduce investors' risks. Only once the technology is 'ready' might they fully insist on the rules of trade liberalization, asking the receiving countries to reduce the respective import barriers.

WTO law plays a role in this constellation. On the one hand, the TRIPS Agreement strengthens the position of technology developers, since it opposes national sovereignty and subsequent protectionism over intellectual property rights. This could lead to enhanced technological innovation (and associated technology transfer).[96] Moreover, the most-favoured-nation provision guarantees that certain measures that facilitate technology transfer towards selected countries are extended to all WTO members.[97] On the other hand, the TRIPS Agreement might render the acquisition of technologies more costly, to the detriment of developing countries.[98]

---

92 Green 2005; Vranes 2011.
93 van Asselt *et al.* 2006, 224.
94 Brewer 2004, 11; van Asselt and Biermann 2007, 502.
95 UNFCCC, Art. 4.5.
96 Littleton 2008.
97 Brewer 2008.
98 Littleton 2008.

Preferential trade measures

The last category also highlights the potential for synergy, and consists of unilateral measures that seek to induce climate action in other countries by offering preferential trade terms. Depending on one's view such measures can either be framed negatively (i.e., preferential treatment is conditional upon climate action), but they could also be framed as positive incentives (i.e., additional benefits will be granted if climate action is undertaken).[99] Several commentators have pointed to the potential of the Generalized System of Preferences[100] to differentiate between developing countries on the basis of climate action if the conditions of the so-called Enabling Clause are met. Based on WTO jurisprudence,[101] this means that such preferential treatment should effectively meet the development, financial and trade needs of developing countries, and that similarly situated countries should receive identical treatment. Given the potential impacts of climate change on developing countries, preferential treatment linked to climate action would arguably meet such needs.[102] Moreover, if the above-mentioned needs are explicitly constructed to refer to addressing climate change, there is an argument that it is possible to grant preferential treatment to countries particularly vulnerable to climate change.[103]

3.2 Border Adjustment Measures

Competitiveness, free rider and carbon leakage concerns

As international efforts to address climate change have intensified, three interrelated problems have become central in climate policy discussions in developed countries: international competitiveness, free riders and carbon leakage. With regard to competitiveness, concerns have been raised that trade-exposed energy-intensive industries will be adversely affected by climate policies compared with their international competitors that operate in countries that do not have binding emission reduction

---

[99] Epps and Green 2010, 182.

[100] McKenzie 2008; Epps and Green 2010, 180–188. The Generalized System of Preferences originated from the United Nations Conference on Trade and Development in 1964 and allows for the preferential tariff treatment of developing countries. The system was endorsed by the parties to the GATT in 1979 through the Enabling Clause.

[101] *European Communities – Conditions for the Granting of Tariff Preferences to Developing Countries*, para. 173.

[102] McKenzie 2008, 691.

[103] McKenzie 2008, 693.

targets in place or that have otherwise less stringent climate policies.[104] A related concern is the free rider problem:[105] if not all countries participate in, or comply with, an international climate change agreement (or adopt climate policies that otherwise impose costs on domestic production), the result is that the goods produced in non-participant countries have a competitive edge over products from countries that do have climate policies in place.[106] Yet another concern, carbon leakage, refers to an increase of emissions in countries without climate policies that can be related to climate policies taken in other countries. Carbon leakage can occur through, among other things, a shift of consumption of carbon-intensive goods towards cheaper import substitutes, resulting in more carbon-intensive production in other jurisdictions or through relocation of industrial production to regions without a carbon price.

Various policy measures to address the concerns arising from the implementation of carbon pricing policies are available, including: (1) measures that lower the carbon costs, such as the free allocation of tradable emission allowances, expanding the scope and coverage of an emissions trading system, or state aid for affected sectors; (2) measures at the border that level the international playing field in either the exporting country with a lower carbon price or in the importing country with a higher price; or (3) measures creating a similar carbon price through the conclusion of international (sectoral) agreements.[107]

Each of these measures has its own advantages and drawbacks in terms of economic and environmental effects, political feasibility and legal acceptability.[108] Since about 2005, much focus has been on the use of

---

104 van Asselt and Biermann 2007, 498.

105 The free rider problem has most famously been described in the context of environmental problems by Garrett Hardin (Hardin 1968, 1244). It has received significant attention from economists and political scientists in the context of international climate policy. Tackling climate change is a global public good, meaning that actors cannot be prevented from enjoying it (it is 'non-excludable'), and that use of the good by one actor does not reduce the amount for other actors (it is 'non-rivalrous'). This leads to particular risks of free riding in international cooperation on climate change, as 'each country can claim for itself only a small fraction of the global benefit of its mitigation efforts, and because marginal abatement costs are increasing', which means that 'the incentive for countries to mitigate climate change on their own is greatly reduced'. Barrett and Stavins 2003, 350

106 Brewer 2003, 331.

107 Neuhoff 2011, 286.

108 Dröge 2009.

border adjustment measures (option 2), essentially a unilateral climate-related trade measure. Two broad design options are commonly distinguished: border tax adjustments, and the requirement for importers to surrender allowances at the border.[109] A border tax adjustment requires importers to pay a charge equivalent to a tax applied to goods produced domestically, whereas a requirement to surrender allowances is linked to an emissions trading scheme and permits goods to enter a country only if a certain amount of emission allowances are purchased that reflect the greenhouse gases emitted during the production process of the good.

Beyond these two broad options, more specific design choices need to be made.[110] First, BAMs can apply to energy-intensive primary goods (e.g., cement, iron and steel, aluminium) or to finished items for consumption (e.g., automobiles, electronic appliances). Second, the country coverage may vary. A measure could apply uniformly to products from all trading partners or to a select group of countries (e.g., those countries that have not undertaken specific climate change mitigation actions). Third, there are different ways of calculating the level of the adjustment (e.g., product-specific, based on a product's carbon footprint; or standardized, based, for instance, on a sectoral average). Finally, there are choices about the timing of a BAM: the measure can come into effect at the same time as the domestic measure or it may be delayed.

Although no such measures have yet been adopted, they have been considered in the US and the EU.[111] It is likely that pressure for their adoption will persist, even under a post-2020 climate agreement. First, there may be free riders under a new agreement, either because some countries do not participate or because the commitments some countries sign up to are perceived as weak by other states. Second, countries or industrial sectors in countries may not comply with their commitments under a new agreement.[112] The larger the (perceived) inequalities between countries and sectors in carbon pricing, the greater the pressure to adopt BAMs. If no new climate agreement is reached in 2015, pressure to adopt the measures will likely be even higher.

---

109 Cosbey 2008. Border adjustment measures can also apply to exports. For instance, exporters may receive a rebate at the border, or could be exempted from the obligation to surrender allowances in an emissions trading system. In line with policy proposals in the US and the EU, my focus here is on import-related measures.

110 Persson 2010, 5–12.

111 Persson 2010, 3–4; van Asselt and Brewer 2010.

112 Holmes *et al.* 2011, 887.

Border adjustment measures and the potential for conflict

The adoption of border adjustment measures would undoubtedly have consequences for the relationship between the trade and climate regimes. While proponents emphasize the need to include such measures to address the triad of competitiveness, carbon leakage and free rider concerns, others have argued that these measures amount to 'green protectionism' or 'eco-imperialism'.[113] There are indeed several compelling reasons for concern, especially for developing countries, who appear most affected by any measure, if adopted.[114] Developing country reactions to the EU's decision in 2008 to include aviation emissions in the emissions trading system are instructive in this regard, as they show the high sensitivity of unilateral climate-related trade measures that affect developing countries.[115]

Depending on their specific design, BAMs may indeed be incompatible with WTO law. While I do not intend to examine the WTO legality of the measures (something that cannot be conclusively established *ex ante* anyway), several design options may affect a measure's WTO compatibility and thus shape the potential conflict between the climate and trade regimes on this matter.

First, the rationale of the measure matters, in particular with respect to the 'necessary' test under Article XX(b) and the 'relating to' test under Article XX(g) of the GATT.[116] While the environmental rationale of minimizing carbon leakage (and possibly inducing other countries to participate in a future climate treaty) could be seen as a legitimate objective, it is unlikely that judicial bodies of the WTO would seriously consider the economic rationale of safeguarding the international competitiveness of energy-intensive industries as a plausible defence.[117] Explicitly mentioning the environmental rationale for the measure would enhance the chances of WTO compatibility. It is unlikely that the WTO dispute settlement bodies would consider 'hidden' non-environmental

---

113 Khor 2009; Barrett 2010, 20.

114 Yu 2009, 19–24; see also Derksen 2011, 20.

115 In January 2012, Directive 2008/101/EC amending Directive 2003/87/EC so as to include aviation activities in the scheme for greenhouse gas emission allowance trading within the Commmunity, [2009] OJ L8/3, entered into force. It covers emissions from all flights arriving in or departing from the EU (subject to limited exemptions) and effectively governs emissions outside of EU airspace. The inclusion provoked strong reactions from developed countries (e.g., the US) and developing countries (e.g., China and India). See Kulovesi 2011a; Scott and Rajamani 2012.

116 Epps and Green 2010, 144–147; Tamiotti 2011, 1207–1208.

117 Pauwelyn 2007, 3–4; Zane 2011, 217–218.

rationales as long as there is a rational connection between the measure and the environmental objective.[118] Having said that, the extent to which the use of the measure effectively contributes to the objective of climate change mitigation can be questioned.[119] For Article XX(b) purposes, it is important to determine whether the measure is 'apt' to make a 'material contribution to the achievement of its objective'.[120] However, several studies cast doubt on the effectiveness of BAMs in achieving their stated goals. Carolyn Fischer and Alan Fox conclude that, although they may have some impact on leakage, an import requirement is unlikely to be very effective as it 'only affects the relative price of domestic and foreign goods in the home country'.[121] Julia Reinaud emphasizes the importance of design details in determining the extent to which border adjustments address leakage and competitiveness, concluding that 'unless [border adjustment] schemes are set on a plant-by-plant basis, rebate exports and cover indirect emissions, it is not clear that they would level the CO_2 playing field to the point where competitiveness levels are restored and leakage is avoided'.[122] In other words, the environmental (and economic) effectiveness of BAMs depends to a large extent on the coverage of goods and sectors, the inclusion of marginal climate policy costs, and the application to all trade flows (imports and exports). In addition, there are practical barriers that may reduce the effectiveness of BAMs.[123] Christopher Weber and Glen Peters show that the primary goods that would most likely be subject to BAMs in the US (e.g., steel, cement, paper) represent a relatively small share of emissions embodied in imports, whereas calculating the emissions embodied in final goods (e.g., electronic appliances, cars) is incredibly complex.[124] These goods are often assembled from parts stemming from different countries, which would require complex measures tracing the emissions in the production process. Trevor Houser and colleagues further demonstrate that BAMs may be ineffective if they target countries from which only small percentages of primary goods are imported, using the example of the US and China.[125] Lastly, the effectiveness of a BAM may be undermined by

---

[118] Howse and Eliason 2009, 72.
[119] Bordoff 2009, 18; Epps and Green 2010, 148–151.
[120] *Brazil – Measures Affecting Imports of Retreaded Tyres*, para. 151.
[121] Fischer and Fox 2009, 25.
[122] Reinaud 2008, 96.
[123] Persson 2010.
[124] Weber and Peters 2009.
[125] Houser *et al.* 2008, 45.

gaming strategies of affected firms and countries, such as attempts to channel goods through countries that are not targeted.[126]

Second, the timing of the measures is important: delaying the effective implementation would buy some time for potentially affected countries to develop and implement domestic climate change mitigation policies and could allow international climate change negotiations to produce results.[127] Demanding that a BAM should be implemented simultaneously with the start of a domestic emissions trading system, as some have argued in the US, would effectively leave no time at all for good faith and serious negotiations, and would undermine the chances of justifying the measure under Article XX.[128]

Third, for determining whether a measure is consistent with the national treatment provision under Article III of the GATT, as well as for its justifiability under Article XX, it matters how the level of the adjustment is determined and whether the measure is applied across the board. In particular, it is important to assess how well the level of the adjustment corresponds to a product's actual emissions.[129] Some of the bills proposed in the US Congress required that all importers of covered goods from covered countries needed to purchase allowances, even if those goods were produced with clean technologies.[130] This would not only discriminate against exporters of low-emission products but also run counter to the environmental objective pursued by the measure. However, tracing the emissions of every product crossing the border poses tremendous practical difficulties.[131] An option would be to use emission standards based on either the carbon content of domestic products or that of imported products.[132] The latter has been advocated by Roland Ismer and Karsten Neuhoff, who suggest that the standard should be the 'best available technology' in the exporting country.[133] Although standards will make the administrative tasks less complex, their application across the board may discriminate against individual exporters that produce goods with lower emissions. A solution would be to allow individual firms to prove that their emissions are lower.[134]

---

126 Houser *et al.* 2008, 56.
127 Haverkamp 2008.
128 Pauwelyn 2007, 40.
129 Tamiotti 2011, 1208.
130 Bordoff 2009, 52–53.
131 Persson 2010, 8–10.
132 Persson 2010, 10.
133 Ismer and Neuhoff 2007, 147.
134 Persson 2010, 6.

Fourth, for the justification of a measure under the chapeau of Article XX, it is of importance how the country implementing the BAM accounts for climate policies in other countries.[135] The proposed 2008 Climate Security Act in the US included a stringent 'comparable action' test, effectively requiring any foreign country to adopt the same kind of emission caps as the US. This could mean that developing countries have to implement and put in place more ambitious climate policies to avoid the imposition of the BAM. The shrimp-turtle cases allowed members to condition market access on the putting in place of regulatory programmes 'comparable in effectiveness' to the programme in the country adopting the measure, as long as 'sufficient latitude … to achieve the level of effectiveness required' is provided.[136] Although most US bills did not directly demand that developing countries take on similar emission reduction targets, and exempted countries with low levels of development and with low emissions (i.e., the least-developed countries), some of them left the determination of 'comparable action' up to US administrative bodies. For WTO compliance, it is important that other countries are involved in the operationalization of the measure.[137] This means that they would need to have some input in the determination of 'comparable action' and that there should be an appeal mechanism in case of disagreement.[138]

Finally, and related to the previous point, whether the measures differentiate between countries is of major importance, most obviously with respect to the most-favoured-nation treatment rule of Article I of the GATT. If a measure is applied to 'like' products based on their country of origin, favouring products from countries with more ambitious climate policies and penalizing products from countries with weak or no climate policies, a violation of this principle is possible. This violation may be avoided by uniformly imposing a BAM on all imported products.[139] However, one environmental rationale for the measure, to induce states to adopt climate measures of their own or join international efforts, would be undermined, with further reverberations when, for instance, the justification of a trade measure under Article XX is ascertained. Moreover, differentiation of financial burdens of climate change mitigation

---

[135] Hertel 2011, 673.

[136] *United States – Import Prohibition of Certain Shrimp and Shrimp Products, Recourse to Art. 21.5 DSU*, para. 144.

[137] *United States – Import Prohibition of Certain Shrimp and Shrimp Products*, paras. 181–183.

[138] Hufbauer *et al.* 2009, 86.

[139] Pauwelyn 2007, 32.

measures between developed countries and developing countries, combined with developed country leadership, is part and parcel of the Kyoto system.[140] This leads to a 'catch-22', where measures that are most consistent with the most-favoured-nation treatment rule are most likely to go against the idea of common but differentiated responsibilities, while measures that differentiate on the basis of climate action undertaken will most likely be viewed as discriminatory under WTO law.[141]

This catch-22 is related to the different ways in which the climate and trade regimes differentiate between countries.[142] Whereas the differentiation in the climate regime is based on the responsibility for causing the problem of climate change (as well as the capacity to deal with the problem), differentiation in the trade regime through 'special and differential treatment' is based primarily on the need to promote trade between countries by making developing countries full partners in the world trading system by treating them favourably.[143] Viewed in this light, BAMs constitute one arena in which developed and developing countries battle over the appropriate level of responsibility for the climate problem.[144] More specifically, the notion of 'comparable action' operationalizes what the country adopting the measure views as the appropriate allocation of responsibilities.[145] A country's perception of comparability could in turn be informed by either the discussions in the climate regime or by WTO jurisprudence.[146] From the perspective of common but differentiated responsibilities and respective capabilities, it is doubtful whether developed countries can actually require developing countries to adopt programmes that lead to a comparable mitigation effect (i.e., emission reductions)[147] or ask for similar technology or regulatory standards.[148] Still, a WTO Dispute Settlement Body is likely to use this standard as long as parties to the climate regime fail to provide clarity about how developed and developing country mitigation efforts could and should be compared.[149]

---

140 Charnovitz 2010b, 408–410; Eckersley 2010, 379–383; Hertel 2011.
141 Eckersley 2010, 382–383.
142 Eckersley 2009, 13.
143 Honkonen 2009, 66–67; see also Pauwelyn 2013.
144 Cosbey 2009, 23.
145 Brewster 2011; Hertel 2011, 677.
146 Werksman 2009b.
147 Hertel 2011, 677.
148 Werksman and Houser 2008, 4.
149 Werksman 2009b, 31.

4. INTERACTION MANAGEMENT: LEGAL TECHNIQUES

4.1 Conflict Avoidance Techniques

Treaty changes and drafting

Similar to the climate–biodiversity case, the first avenue for avoiding incompatibility between the climate treaties and the WTO agreements would be through careful drafting of a new treaty or amending the existing treaties.[150] I have noted that climate negotiators have been reluctant to explicitly address trade issues beyond the general provisions of the climate treaties. However, some countries have become more outspoken since the mid-2000s.

The potential trade impacts of climate policies were initially only discussed in negotiations related to the impact of the implementation of response measures.[151] Following the Bali Action Plan, climate negotiators also started to address the issue of trade measures under the heading of 'economic and social consequences of response measures' in the AWG-LCA.[152] In the climate negotiations in 2009, India, supported by the G77 and China,[153] proposed that: '[d]eveloped country Parties *shall not resort to any form of unilateral measures*, including fiscal and nonfiscal border measures against goods and services imported from other Parties, in particular from developing country Parties, on grounds of stabilization and mitigation of climate change'.[154]

In the negotiation text under discussion at the Cancún COP in 2010, four alternatives were outlined. One of these[155] reiterated the position of India, adding that such measures contravene the provisions and principles of the UNFCCC, particularly that of common but differentiated responsibilities[156] and the link between developing country mitigation action and the provision of financial and technological support by developed countries.[157] Other proposals, however, largely amount to a reiteration of Article 3.5 of the UNFCCC.[158] Parties could not agree on a specific text,

---

150 Eckersley 2010, 387–388; Epps and Green 2010, 228–233.
151 UNFCCC, Art. 4.8; Kyoto Protocol, Art. 2.3 and Art. 3.14.
152 UNFCCC Decision 1/CP.13, para. 1(b)(vi).
153 Hertel 2011, 658.
154 Third World Network 2009 (emphasis added).
155 UNFCCC 2010c, 26.
156 UNFCCC, Art. 3.1.
157 UNFCCC, Art. 4.7.
158 UNFCCC 2010c, 26–27.

but rather requested the subsidiary bodies to convene a forum on the impact of the implementation of response measures, which could in turn result in 'a possible forum on response measures'.[159] The forum on the impact of the implementation of response measures was used by some parties and observers to highlight the socio-economic impacts of climate policies on trade,[160] indicating that the issue will remain under discussion.[161] Even if some form of agreement on the text is possible, however, it is far from certain that such a provision will make it into a new, legally binding agreement.

From the WTO perspective, one suggestion is to amend Article XX of the GATT to explicitly accommodate climate-motivated trade measures. The challenges to this approach are primarily related to political feasibility. First, WTO members would need to trust that the WTO adjudicators can and will distinguish between legitimate and illegitimate BAMs. Second, a broad standard comparable to the 'necessity' and 'related to' tests needs to be agreed upon.[162] Other, less formal options for accommodating BAMs in the WTO are conceivable and will be discussed under the heading of institutional coordination.

Like the climate–biodiversity case, the application of this legal technique, either with respect to the climate or the trade agreements, is hence limited by what is politically feasible.

Treaty interpretation

Treaty interpretation as a technique for avoiding conflict between the climate and trade regimes has been discussed at great length in the literature.[163] Although it is possible that WTO members themselves issue an interpretative statement on the relationship between the climate and trade treaties,[164] treaty interpretation is likely to take place before the WTO's Dispute Settlement Body.

Article 31.3(c) of the VCLT can be seen as a promising avenue for avoiding a conflict between the climate and trade treaties (see Chapter 5). To Campbell McLachlan, supported by the ILC, the interpretative guidance contained in this provision amounts to a 'principle of systemic

---

159 UNFCCC Decision 1/CP.16, paras. 88–94.

160 http://unfccc.int/cooperation_support/response_measures/items/6009.php (accessed 11 December 2013).

161 International Centre for Trade and Sustainable Development 2011.

162 Epps and Green 2010, 259–260.

163 Notably Voigt 2008a, 265–292.

164 Stokke 2004, 351–352.

integration',[165] whereas Riccardo Pavoni views it as a 'principle of mutual supportiveness' governing the relationship between the WTO and MEAs.[166]

The notion of systemic integration, if used more consistently by treaty interpreters, could effectively allow the dispute settlement bodies of the trade regime to rely on norms originating in other international regimes. In a dispute before the WTO on the domestic application of a climate-related trade measure, such as a BAM, Article 31.3(c) could play a role if international climate change law were to be used to interpret obscure WTO norms. Before I discuss the possibilities of doing so under WTO law and practice, it is useful to examine the circumstances under which the WTO dispute settlement bodies could seek recourse to international climate change law in case of a BAM-related dispute. Several possibilities come to mind, mostly in relation to Article XX of the GATT.[167]

First, climate treaties, and in particular a possible future climate agreement, could inform the analysis under Article III of the GATT. In particular, they could show that 'higher taxation of unlike but directly competitive products is not applied for protectionist reasons' but with a view to reducing global emissions.[168] The climate treaties *ceteris paribus* strengthen the case of parties adopting measures against non-parties. However, whether they also strengthen the case of parties adopting the measure against other parties is questionable and would depend on various factors, including the level of mitigation action required by the country adopting the measure vis-à-vis the country affected by it, and the legal nature of commitments. The usefulness of a post-2020 climate agreement may be reduced given its likely shift towards self-selected commitments (see Chapter 2).

Second, under the analysis of Article XX of the GATT, the climate treaties could assist in interpreting whether a BAM is '*necessary* to protect human, animal or plant life or health',[169] or '*relating to* the

---

[165] McLachlan 2005, 280; ILC 2006, paras. 410–480.

[166] Pavoni 2010, 678.

[167] This focus on Article XX presupposes that the BAM will violate the non-discrimination provisions of the WTO. This does not have to be the case. Ismer and Neuhoff argue that their proposed design would not lead to such a violation, making recourse to Article XX unnecessary. See Ismer and Neuhoff 2007, 149. However, most proposed design options would require a justification through Article XX, particularly if the BAMs differentiate between countries on the basis of 'comparable action' or ratification of a climate treaty.

[168] Wiers 2008, 28.

[169] GATT, Art. XX(b) (emphasis added).

conservation of exhaustible natural resources'.[170] A country adopting the measure may invoke its ratification of climate treaties in its defence of the non-commercial, environmental objectives of its measure, for instance by arguing that the climate treaties underline the importance of the objective of climate change mitigation. Again, all other things being equal, participation in climate treaties would make the tests formulated in Article XX (b) and (g) generally easier to meet.

Third, the chapeau of Article XX constitutes one notoriously ambiguous provision. Recourse to the UNFCCC in this regard would be particularly interesting, as it contains a nearly-verbatim provision.[171] However, it should not simply be assumed that these provisions have the same meaning: WTO jurisprudence on this phrase has developed in a WTO context, whereas the UNFCCC provision needs to be read within its UNFCCC context.[172] What this could mean is that measures tailored to countries' responsibility for causing the climate change problem and their capacity to mitigate would stand a stronger chance of being justified under Article XX than measures that are uniformly applied to all countries,[173] although, as noted above, this would mean that the measure violates the most-favoured-nation rule.

Finally, the climate regime may be drawn on indirectly to show compliance with criteria from past WTO jurisprudence. For instance, as interpreted by the WTO's dispute settlement bodies, the chapeau of Article XX contains a 'prior negotiation efforts' criterion, according to which a unilateral measure will not be considered discriminatory if it has been preceded by serious bilateral or multilateral negotiation efforts.[174] According to Joost Pauwelyn, this requires 'at the very least good faith efforts … to bring … countries into the fold of an international effort to combat climate change before making a move to the second or third best option of unilateral border adjustments'.[175] I would add that this would also require more detailed negotiations with potentially affected countries on the design and application of the measure.

These examples illustrate the possible role of the climate treaties (or negotiations) in WTO dispute settlement and potentially in reconciling climate and trade objectives. But when and how would WTO dispute

---

170 GATT, Art. XX(g) (emphasis added).

171 UNFCCC, Art. 3.5.

172 Hertel 2011, 663.

173 Hertel 2011, 677.

174 *United States – Import Prohibition of Certain Shrimp and Shrimp Products*, para. 166.

175 Pauwelyn 2007, 40.

settlement bodies use non-WTO law? And, compared to other treaty interpretation methods, how can Article 31.3(c) be of particular use?

The question on the WTO dispute settlement bodies' use of non-WTO law in general and international environmental law in particular has received much attention,[176] mainly because there is relevant case law. Three ways in which the WTO adjudicators can use non-WTO law (e.g., the climate treaties) can be distinguished: (1) as applicable law between the parties in the dispute; (2) as an interpretative device of WTO terms; or (3) as factual evidence.[177]

The issue of applicable law has been the most controversial. If non-WTO law could be used as applicable law, this would mean that the climate treaties could be used as an independent defence of the violation of WTO norms, a position that has been prominently defended by Pauwelyn,[178] but countered by Gabrielle Marceau and Joel Trachtman.[179] Referring to the *Korea-Government Procurement* ruling,[180] Pauwelyn argues that non-WTO law automatically applies.[181] Counter-arguments are related to fears of intrusion of non-WTO law[182] and concerns about WTO adjudicators getting involved in lawmaking activities.[183] So far, the WTO dispute settlement bodies have refrained from directly applying non-WTO law, mainly because there was no perceived need to do so.[184]

Even if non-WTO law cannot be directly relied upon, treaty interpretation could open the door. One of the often-used examples of the WTO's openness to non-WTO law is the *US-Shrimp* case, in which the Appellate Body referred to international environmental documents such as Agenda 21, UNCLOS, the CBD and the Convention on the Conservation of Migratory Species of Wild Animals to interpret 'exhaustible natural resources' under Article XX(g),[185] and to the Rio Declaration, Agenda 21, the CBD and the Convention on the Conservation of Migratory

---

[176] Marceau 2001; Trachtman 2002; Pauwelyn 2003a; Pauwelyn 2003b; Kulovesi 2011b.

[177] M.A. Young 2011b, 195–206.

[178] Pauwelyn 2003a, 460 and 466; Pauwelyn 2003b, 997–998.

[179] Marceau 1999, 110; Marceau 2001, 1116; Trachtman 1999, 342.

[180] *Korea – Measures Affecting Government Procurement*, para. 7.96.

[181] Pauwelyn 2003, 466.

[182] Voigt 2008a, 285.

[183] Cossy and Marceau 2009, 385.

[184] E.g., M.A. Young 2011b, 196.

[185] *United States – Import Prohibition of Certain Shrimp and Shrimp Products*, para. 130.

Species with respect to the interpretation of the chapeau of Article XX.[186] It is unclear, however, whether the Appellate Body referred to these documents because of their legal status under Article 31.3(c) or as information to find out the 'ordinary meaning' of terms under Article 31.1 of the VCLT.[187]

Indeed, while Article 31.3(c) of the VCLT has been invoked by various international adjudicatory bodies,[188] the WTO dispute settlement bodies have been reluctant to seek recourse to it,[189] and it is far from certain whether they would do so in a future climate–trade dispute. In particular in the *EC-Biotech* dispute, the WTO Panel was very cautious in using provisions from other international treaties to interpret WTO norms, in this case the CBD and the Cartagena Protocol on Biosafety.[190] The Panel seemed to suggest that 'the parties' referred to in Article 31.3(c) should be all the parties to a treaty being interpreted (i.e., the Agreement Establishing the WTO), rather than, for instance, the parties to the dispute.[191]

The Panel's finding has been heavily criticized by the ILC (as well as other scholars), who have pointed out that this amounts to a de facto requirement of parallel membership, meaning that non-WTO law, including the climate treaties, would hardly have any role to play in WTO dispute settlement.[192] However, the *EC-Biotech* Panel continued to state that non-WTO law could still inform the 'ordinary meaning' of terms under Article 31.1.[193] Under this approach, the climate treaties (and decisions adopted under them) could fulfil a similar role as a dictionary, and could still be of use even if treaty membership is not congruent. Still, despite their obvious relevance to the dispute, it is notable that the Panel did not seek recourse to the provisions of the CBD or the Cartagena Protocol in the context of Article 31.1, although it did use various 'conventions, standards and guidelines ... as appropriate'.[194] This indirect

---

[186] *United States – Import Prohibition of Certain Shrimp and Shrimp Products*, para. 168.

[187] Pauwelyn 2003, 256; M.A. Young 2011b, 199.

[188] ILC 2006, paras. 433–460.

[189] Lindroos and Mehling 2008; Van Damme 2009.

[190] *European Communities – Measures Affecting the Approval and Marketing of Biotech Products*, paras 7.68–7.72.

[191] In defence, see Linderfalk 2008.

[192] ILC 2006, para. 450; see also M.A. Young 2007.

[193] *European Communities – Measures Affecting the Approval and Marketing of Biotech Products*, paras. 7.90–7.96.

[194] *European Communities – Measures Affecting the Approval and Marketing of Biotech Products*, para. 7.96.

way of taking into account other treaties, which neither clarifies when non-WTO law needs to be taken into account, nor how this should be done, was aptly characterized by the ILC as 'rather contrived'.[195]

Another way in which treaty interpretation could take into account non-WTO law would be through a teleological interpretation of the provisions of the WTO agreements, taking into account their 'context' and their 'object and purpose'.[196] This interpretation method could explicitly include the preambular language on the WTO's sustainable development objective. While such an interpretation does not make protectionist measures WTO compatible,[197] it could allow for a balancing approach that would not be too different from the one envisaged under Article 31.3(c).[198] However, as the Panel ruling in the *US-Shrimp* case showed, a teleological approach does not always lead to a balanced discussion, as the Panel set aside the sustainable development objective, arguing that the central focus of the Agreement Establishing the WTO 'remains the promotion of economic development through trade'.[199]

Furthermore, the WTO dispute settlement bodies may adopt an 'evolutionary approach' to interpretation of the WTO agreements irrespective of their reference to Article 31.3(c). Although Pauwelyn discusses the notion of 'evolutionary interpretation' mainly in the context of Article 31.3(c),[200] it can equally be combined with the other rules of interpretation of Article 31.[201]

Lastly, non-WTO law could play a role as relevant facts (or evidence of such facts).[202] This approach was also adopted in the *US-Shrimp* case, where the Appellate Body used the fact that sea turtles were listed in the Convention on International Trade in Endangered Species of Wild Fauna and Flora as evidence of *the fact* that they were threatened with extinction (thus supporting *the interpretation* that they were 'exhaustible').[203] In a similar way, the WTO dispute settlement bodies could use the climate treaties to establish that certain measures de facto contribute

---

195 ILC 2006, para. 450.

196 VCLT, Art. 31.1.

197 Voigt 2008a, 279.

198 See also M.A. Young 2011b, 234, 236–237.

199 *United States – Import Prohibition of Certain Shrimp and Shrimp Products*, Report of the Panel, para. 7.42.

200 Pauwelyn 2003a, 264–268.

201 Voigt 2008a, 275–276.

202 M.A. Young 2011b, 204–206.

203 *United States – Import Prohibition of Certain Shrimp and Shrimp Products*, para. 132.

to emission reductions, although IPCC reports or similar credible scientific sources would likely form more appropriate source material.

This discussion shows that there is a variety of ways in which climate change law could be taken into account in a WTO dispute (with a diminishing degree of intrusion): directly, through applicable law; indirectly through treaty interpretation under either Article 31.3(c) or Article 31.1 of the VCLT; or as factual evidence.

Returning to the usefulness of the principle of systemic integration, it is unclear whether Article 31.3(c) is able to prevent conflicts between the climate and trade regimes. First of all, it remains unsettled what 'taken into account' actually entails in the context of WTO dispute settlement.[204] It is generally agreed that this phrase does not mean that the extraneous rules override the interpreted rules, but rather that their normative significance needs to be determined on a case-by-case basis.[205] So while the legal norms in the climate regime could inform a decision by the WTO dispute settlement bodies, treaty interpretation as such cannot set aside WTO norms.[206] This can only be achieved if one follows Pauwelyn's proposal to directly apply non-WTO law in a dispute.

Second, the predominant approach adopted in WTO jurisprudence so far is that Article 31.3(c) has mainly 'residual value',[207] and that international environmental law is mainly relied upon to inform the 'ordinary meaning' under Article 31.1, or as factual evidence. This situation can be criticized: as Kati Kulovesi convincingly argues, the WTO's approach to international environmental law is inconsistent, unpredictable and biased against international environmental law to the extent that the latter does not contribute to the trade regime's goals of market liberalization.[208]

A more consistent and predictable balancing approach could arguably be based on the principle of systemic integration.[209] However, although the principle can be regarded as a strong integrative device in theory, this

[204] Sands 1999, 103; ILC 2006, para. 423.

[205] ILC 2006, para. 473–474; Voigt 2008a, 282.

[206] Voigt 2008a, 284–286.

[207] Van Damme 2009, 375; Samson 2011, 711–712.

[208] Kulovesi 2011b, 153–178. Kulovesi shows, for instance, how the Appellate Body has been keen to refer to Principle 12 of the Rio Declaration (on the relationship between trade and the environment), but has been reluctant to use the precautionary principle, even though the status of the latter is less disputed in international environmental law. This is evidence, she argues, of the WTO's 'selective incorporation' of international environmental law. See also Dunoff 1998, 24–25.

[209] E.g., Matz-Lück 2012, 229.

theoretical strength is its weakness in judicial practice. Adjudicators will refrain from using it, as the resulting normative integration would also entail integration of authority (i.e., a direct influence on lawmaking).[210] As Broude explains, 'to integrate (with) the norms of another system is to acknowledge the authority of that other system to produce pertinent norms' as well as 'assert[ing] authority over them'.[211] It is this authority integration that dispute settlement bodies seek to avoid by using 'weaker' forms of integration, such as Article 31.1 of the VCLT, or by simply using non-WTO law as factual evidence. Similarly, while a possible principle of mutual supportiveness between WTO law and MEAs has much theoretical appeal as an interpretative device in a possible climate–trade dispute,[212] its use in judicial practice cannot be assumed. This is all the more unlikely given that, unlike various other MEAs, the climate treaties do not contain language (either in their preambles or in their operative parts) stating explicitly that trade and environmental policies should be mutually supportive.

4.2 Conflict Resolution Techniques

The tensions related to BAMs are best classified as policy conflicts, as the provisions of the UNFCCC and the Kyoto Protocol provisions do not suggest WTO-incompatible behaviour and leave the design and use of BAMs up to their parties. This situation would change quite substantially if parties to the climate regime attempted to incorporate such measures in a future climate agreement through novel treaty provisions or by means of a COP decision.[213]

At the moment, a scenario in which parties to the UNFCCC adopt multilateral trade measures mandating BAMs against other (non-complying) parties may be 'extremely unlikely',[214] but it cannot be ruled out completely. In comparison, a provision or decision obliging countries to adopt BAMs aimed at non-parties to a future agreement may be more feasible. A precedent for multilateral trade measures exists under the Montreal Protocol, which bans the trade in ozone-depleting substances

---

[210] Broude 2008, 200.

[211] Broude 2008, 186–187.

[212] Pavoni 2010.

[213] UNFCCC, Art. 4.2(d), Art. 7.2(e) and Art. 13.4(a), allow the COP to validate an indirect international climate measure. Therefore, the COP decision itself, which validates the domestic climate measure, can conflict with a WTO norm.

[214] Cosbey 2009, 24.

with non-parties,[215] a blunter instrument compared to BAMs.[216] While the Montreal Protocol analogy may not be entirely appropriate for the climate regime,[217] it does show that multilateral trade measures are conceivable in MEAs.[218] If new treaty language or a COP decision were to be adopted to this end, the relationship between norms of the climate and trade regimes would more likely fall under the definition of normative conflict outlined by scholars such as Pauwelyn and Erich Vranes (see Chapter 4).

Conflict clauses

None of the treaties in question contain any clearly identifiable conflict clauses.[219] In the negotiations leading up to the Kyoto Protocol, Australia proposed inserting a provision that would have explicitly prevented derogation from the WTO (among other international agreements), but this proposal did not make it into the final text.[220]

Article 3.5 of the UNFCCC and Article 2.3 of the Kyoto Protocol could be seen as conflict clauses. However, these provisions do not establish a clear relationship between the two regimes, and do not allow or disallow climate-related trade measures.[221] In other words, they do not determine which treaty would prevail in the case of a conflict arising. Still, it is important to also note what the agreements do *not* explicitly state: they do not subordinate to the WTO agreements, in contrast to, for instance, the Cartagena Protocol on Biosafety; and they do not explicitly allow for trade measures against non-parties or non-compliers.[222] Furthermore, while not being explicit conflict clauses, the provisions may still provide interpretative guidance. For instance, Christina Voigt argues that Article 2.3 of the Kyoto Protocol directs parties to adopt measures that minimize effects on international trade, except in cases where such effects are necessary to ensure the effectiveness of such measures.[223] However, given the different possible rationales for BAMs (protecting competitiveness, addressing carbon leakage and tackling the free rider

---

215 Montreal Protocol, Art. 4.
216 Barrett 2010, 10.
217 Cosbey 2009, 22–24.
218 Barrett 2010.
219 Voigt 2008a, 293–299.
220 Depledge 2000, paras. 500–501.
221 Voigt 2008a, 299.
222 Stokke 2004, 346.
223 Voigt 2008a, 298.

problem), I would question whether it is possible to establish what the intended effects are in the first place.

The WTO agreements also do not contain any explicit conflict clauses relating to the climate treaties.[224] The Agreement Establishing the WTO does not contain a conflict clause clarifying how it relates to pre-existing international law, including the UNFCCC.[225] While the mandate of the CTE includes an examination of the relationship between the WTO and MEAs, the CTE has not specified what should happen if a conflict arises. The CTE also has not followed up on its mandate allowing it to propose modifications to the multilateral trading system.[226]

Priority rules

For the conflict resolution techniques of Article 30 of the VCLT to be applicable, it must first be established that the climate and trade treaties relate to the 'same subject matter'.[227] According to Pauwelyn, two treaties must relate to the same subject matter as otherwise there would be no conflict in the first place.[228] This implies that the applicability of Article 30 depends on the existence of a conflict of norms. As noted above, there is no such conflict as of yet between the climate and trade treaties, although this may change over time. The ILC takes a broader view of the phrase, stating that whenever 'the fulfilment of the obligation under one treaty affects the fulfilment of the obligation of another', it can be said that two treaties relate to the same subject matter.[229] This could mean that conflicts between the climate and trade regimes could potentially be resolved through the tools offered by the Vienna Convention. However, other commentators have argued for a more stringent interpretation of the article, arguing that 'the [VCLT] is not applicable to the thornier issues of what happens when treaties have different foci but overlapping issue areas'.[230]

Even if we accept that Article 30 is applicable, it is difficult to see how its conflict resolution techniques could be useful. The *lex posterior* rule codified in the Vienna Convention provides guidance on prioritizing successive treaties in theory, but would be difficult to apply in practice. Voigt explains that, according to the maxim, the Kyoto Protocol (adopted

224 Voigt 2008a, 295.
225 Pauwelyn 2003a, 343–344.
226 WTO 1994.
227 VCLT, Art. 30.
228 Pauwelyn 2003a, 364.
229 ILC 2006, 130.
230 Borgen 2005, 603.

in 1997) and a possible future climate treaty would prevail over the WTO agreements (adopted in 1994), while the latter would prevail over those of the UNFCCC (adopted in 1992), 'leading to the odd situation that … the references to the Convention would be cut off'.[231] Furthermore, the nature of the climate treaties as 'living treaties' makes it difficult to pinpoint the latest date of the expression of state consent.[232]

With respect to the *lex specialis* maxim, Olav Schram Stokke argues that 'any explicit provision for climate-related trade measures would have been more specific than the general WTO rules'.[233] This is not necessarily the case, however, as there is no agreement on whether the *lex specialis* rule applies to a specific provision or a treaty as a whole.[234] In the latter case it would be particularly difficult to argue that a climate treaty is more specific than a trade agreement (or vice versa). Buck and Verheyen argue that the former application is more appropriate to ensure a mutually supportive relationship between the climate and trade regimes.[235] Still, such a blanket statement on the applicability of the maxim in the climate–trade conflict is premature, as determining which law is more specific depends on one's systematization of international law.[236] Put differently, should BAMs be viewed as climate-related trade measures or as trade-related climate measures?[237] There is no definitive answer. Moreover, it is uncertain whether the WTO dispute settlement bodies would actually apply the *lex specialis* rule in the context of the relationship with non-WTO law; they have been unwilling to do so as of yet.[238]

5. INTERACTION MANAGEMENT: INSTITUTIONAL COORDINATION

5.1 Existing Institutional Coordination

The dispute settlement route is not the only way to address the relationship between the climate and trade regimes with respect to BAMs.

---

231 Voigt 2008a, 303; see also Buck and Verheyen 2001, 34–35.
232 Pauwelyn 2003a, 378.
233 Stokke 2004, 346.
234 Voigt 2008a, 300.
235 Buck and Verheyen 2002, 35.
236 Lindroos 2005, 66.
237 Cf. Michaels and Pauwelyn 2011, 36.
238 Lindroos 2005, 56–58.

Institutional coordination may provide an important alternative means of managing interactions. The record of institutional coordination involving both regimes, however, is limited, even if efforts are intensifying.

The climate and trade regimes provide for institutional coordination with other intergovernmental organizations. The Agreement Establishing the WTO states that the 'General Council shall make appropriate arrangements for effective cooperation with other intergovernmental organizations that have responsibilities related to those of the WTO'.[239] Similarly, the UNFCCC calls on its COP to 'seek and utilize … the services and cooperation of, and information provided by, competent international organizations and intergovernmental and non-governmental bodies'.[240]

A first step in operationalizing these mandates, mutual observership, has been taken. The UNFCCC has had permanent observer status in the CTE since 1999 and is an ad hoc observer at the CTE Special Sessions.[241] The reason for this ad hoc observership is usually explained with reference to a general moratorium in the WTO General Council on granting observer status,[242] but there is also still resistance from countries against permanent observership of certain MEAs in the CTE Special Sessions.[243] The observership in the CTE has meant that the UNFCCC secretariat has regularly participated in CTE meetings,[244] for instance, updating the CTE on developments in the climate regime. The most notable activity of the UNFCCC secretariat has been the drafting of a note summarizing the state of the trade negotiations for participants in the climate regime.[245]

The WTO secretariat, and in particular its Division on Trade and Environment, has become active in enhancing the transparency and raising awareness of the WTO activities related to climate change,[246] for instance through presentations at the climate negotiations. Although the bureaucracies can play a role in making information more widely available and raising awareness about the WTO's approach to the climate–trade overlap,[247] similar limitations to the ones discussed in the

---

239 Agreement Establishing the WTO, Art. V.1.
240 UNFCCC, Art. 7.2(h); see also Kyoto Protocol, Art. 13.4(i).
241 Motaal 2002.
242 Cossy and Marceau 2009, 374.
243 Jinnah 2010, 67–68.
244 WTO 2007.
245 UNFCCC 2003.
246 Jinnah 2010, 68.
247 Jinnah 2010.

previous chapter can be observed. As Mireille Cossy and Gabrielle Marceau argue, 'the competences of the secretariats are limited (they do not normally include decision-making) and underlain by their obligation to remain neutral vis-à-vis the membership'.[248] In other words, interaction management activities of the WTO secretariat are likely to remain limited to exerting cognitive influence.

Cossy and Marceau's comment can be exemplified by the reactions to the 2009 report on climate and trade co-authored by staff from the WTO secretariat and the Economics and Trade Branch of UNEP.[249] Press reports surrounding the release of the report were ambiguous about whether the report was prepared and released by the WTO (and, subsequently, whether it would have any legal effect), or whether it was merely a background report by the secretariat.[250] This ambiguity, as well as some of the findings of the report regarding BAMs, drew strong reactions from some WTO members (e.g., China, India and Mexico) in a CTE meeting following its release.[251] These reactions show that in making statements about the WTO compatibility of climate-related trade measures, the secretariat will need to tread carefully. Other members (e.g., Australia, Brazil, the EU, South Africa, the US and Uruguay), viewed the report as helpful in starting discussions on the topic within the Committee. This does not mean, however, that members are willing to initiate formal discussions on climate and trade within the WTO. As a report from a subsequent meeting indicates, members 'wished to proceed cautiously' and 'to avoid any discussion that could affect the position of countries in the context of the UNFCCC process'.[252] This divide was still clear in 2011, when the CTE reported that although some members were supportive of starting a dialogue on BAMs in the CTE, 'other Members expressed concerns on the appropriateness of the CTE as a forum to discuss climate change-related issues'.[253] This means that while members have paid heightened attention to the topic, substantive discussions, particularly on controversial topics such as BAMs, remain limited.

---

248 Cossy and Marceau 2009, 376.
249 UNEP and WTO 2009.
250 Charnovitz 2010a, 274–275.
251 WTO 2009b, 12–13.
252 WTO 2009b, 6.
253 WTO 2011b, para. 13.

5.2 Enhancing Institutional Coordination

Options for enhancing institutional coordination

Several options for enhanced cooperation and coordination have been proposed in the literature. For instance, it has been suggested that the UNFCCC and WTO secretariats could conclude a non-legally binding Memorandum of Understanding on the use of BAMs by parties to the Kyoto Protocol.[254] Parties to the WTO and the UNFCCC could also establish a consultative mechanism to discuss the adverse effects of climate policies, or they could establish a joint WTO/UNFCCC working group.[255] Enhanced institutional coordination has been discussed in the CTE Special Sessions, resulting in a draft decision text that remains to be adopted. The text provides for more regular information exchange between the WTO and MEA secretariats, and proposes the establishment of a group of experts to provide technical assistance to developing countries in implementing trade measures in MEAs.[256]

However, the feasibility of these options is limited for various reasons. The first is the conflict of ideology and interests between the two regimes: trade liberalization and cheap production processes (WTO) versus environmental protection and sustainable production processes (UNFCCC/Kyoto Protocol). More specifically, Charnovitz posits that:

> [t]he climate regime is driven by the need to correct market failure. Therefore, governments want maximum flexibility at the national level in using economic instruments to influence individual behavior. By contrast, the trade regime is not a response to market failure; it is a response to government failure, that is, the distortions of policy fomented by mercantilism and protectionism.[257]

Second, membership matters. In this case, it is notable that the US is a WTO member but has not ratified the Kyoto Protocol, making the achievement of the required consensus for implementing more formalized institutional coordination challenging in the WTO, even though this may change if and when a new climate agreement 'applicable to all' is adopted in 2015.

Third, those who engage in the negotiations of the two different issues are usually not the same individuals: generally it is representatives of trade ministries that attend WTO discussions and negotiations, whereas

---

[254] van Asselt *et al.* 2005, 262.
[255] Stokke 2004, 351; Assunção and Zhang 2002, 21.
[256] WTO 2011c, Annex.
[257] Charnovitz 2003, 143.

representatives from environment ministries attend UNFCCC negotiations (although this is not necessarily the case for all countries).

While the potential for institutional coordination between the UNFCCC and WTO may thus seem limited, coordination on BAMs may serve several important purposes. I now turn to the question of how such coordination may address several concerns, and identify several issues that could be the subject of coordination. I then discuss how BAMs could potentially be coordinated and disciplined under the UNFCCC, the WTO or in an alternative forum.

Rationales for enhanced institutional coordination on border adjustment measures

The unilateral introduction of BAMs might have negative political implications for both the climate and trade negotiations. In the context of the climate negotiations, applying trade measures as a 'stick' to enforce participation may backfire, leading to entrenched positions by key actors in the climate regime. The prospect of unilateralist behaviour has also sparked fears of trade wars.[258] Indeed, this seems to be the impression conveyed by countries like India in both the climate change and the trade negotiations,[259] and is confirmed by the reactions to the inclusion of aviation in the EU emissions trading scheme. However, if unilateral measures are inevitable (a scenario only becoming more likely if efforts to curb emissions are stepped up), international coordination on the possible application of BAMs could allay some of these fears.

While it may make sense for policymakers in the EU and US to start bilateral discussions on the topic, this would other countries the impression that both intend to go ahead with the option. Applying BAMs will very likely have an impact on countries outside of the EU and the US.[260] Hence, even if the EU and US (and perhaps other developed countries) were to reach agreement, it is conceivable that affected developing countries would seek a different forum to protect their interests.[261] Therefore, it is more sensible to start with broader consultations with countries that are likely to be affected.[262]

There are several rationales for coordination to discipline the unilateral application of BAMs.[263] First, coordination could enhance transparency:

---

[258] Bordoff 2009, 41.
[259] WTO 2009b, 13; UNFCCC 2011b.
[260] Derksen 2011.
[261] Brewster 2011.
[262] Eckersley 2010, 387.
[263] Neuhoff *et al.* 2008, 5; Persson 2010, 13.

it could ensure that potentially affected developing countries are convinced that the measures will not be used to discriminate against their producers. Second, coordination could enhance predictability: it could prevent the emergence of (diverging) systems that would pose an excessive burden for exporters from affected countries. Third, disciplining the use of BAMs could smooth the transition to full carbon pricing policies across the globe while addressing potential leakage problems. Finally, it makes sense to pursue international coordination for more strategic legal reasons: if the EU or the US adopts the measure after all, it would at least provide them with some credit in case a WTO dispute arises.[264]

Coordination could be used to address numerous issues.[265] First, the product coverage could be discussed to provide clarity about which (sub-)sectors are actually exposed to leakage risks, and consequently for which sectors measures may make sense. Second, the country coverage could be discussed. This would include a discussion of how climate policies in different countries could be compared,[266] for example through comparing policies with quantitative and qualitative objectives, and comparing policies with short-term and long-term objectives.[267] Furthermore, exemptions for countries with low emissions and/or low capacity to reduce or limit emissions could be agreed upon.[268] Third, international discussions could seek to agree on how to calculate the border adjustment, for example by using expert bodies to identify the appropriate level of adjustments. Finally, international coordination could go hand-in-hand with broader discussions on how to engage developing countries in future climate action. Rather than simply discussing the 'stick' of trade measures, this could include issues that developing countries perceive to be more important, such as the transfer of clean technologies, REDD and funding for adaptation to climate change.[269] This approach is taken in the Montreal Protocol, in which trade restrictions are combined with financial support and technology transfer.[270]

One way of combining 'carrots' and 'sticks' is through recycling the revenues of the measure, for instance the revenues from auctioning

264 Hufbauer *et al.* 2009, 56–57.
265 Neuhoff *et al.* 2008, 10–11.
266 Hufbauer *et al.* 2009, 108.
267 See also Houser *et al.* 2008, 39.
268 Neuhoff *et al.* 2008, 10.
269 Zhang 2009, 84.
270 Barrett 2010.

allowances under the EU emissions trading system.[271] Indeed, BAMs could become both legally and politically more feasible if they were accompanied by a reimbursement mechanism that leaves the general principle of common but differentiated responsibilities intact.[272] This could be a mechanism that redistributes the proceeds generated through levying duties on importing firms from developing countries back to these countries, either directly or through an international fund. Such a mechanism would simultaneously help achieve the goals of climate protection and a fair international playing field in energy-intensive industries. The most advanced proposal in this regard has been tabled by Michael Grubb, who argues that coupling BAMs with climate finance may be attractive for all parties concerned.[273] For parties wishing to adopt the measures, taking away the fear of carbon leakage could remove the pressure to continue the inefficient system of free allocation of allowances under a domestic emissions trading scheme, while at the same time forming an incentive to increase the ambition of domestic climate policies. Exporting countries targeted by the measures should in theory also welcome such changes in the policy of the importing country. More importantly, however, they would likely welcome the new climate finance.[274] Although the extent to which developing countries would indeed see the recycling of revenues as 'new and additional' climate finance can be questioned, it would at least open up discussion on the extent to which the responsibility for emissions lies with either the producing or the consuming country.

Coordinating border adjustment measures and forum choice

While there are convincing arguments for initiating coordination of BAMs, the most appropriate forum for coordinating BAMs is an open question. Related to this, it remains to be seen to what extent such discussions could be formalized, for example, by means of a multilateral agreement.[275]

---

[271] Cf. Directive 2003/87/EC of the European Parliament and of the Council of 13 October 2003 establishing a scheme for greenhouse gas emission allowance trading, as amended by Directive 2009/29/EC, OJ [2009] L140/63, Article 10.3.

[272] Eckersley 2010, 389.

[273] Grubb 2011.

[274] Grubb 2011, 1055.

[275] Neuhoff *et al.* 2008, 11.

A first option would be to discuss BAMs (or more broadly: climate-related trade measures) in the context of the UNFCCC.[276] Discussions on BAMs would inevitably tie in with the ongoing climate negotiations, in particular with discussions on 'comparability'[277] as well as discussions on response measures. Furthermore, given the need to discuss sticks in conjunction with carrots, there is a compelling argument to take this up in the climate negotiations. However, as noted above, it is unclear whether parties to the UNFCCC would be willing to take up such a sensitive subject. Moreover, it can be questioned whether it is sensible to add more topics to an already overburdened negotiation process.[278] Nevertheless, the insistence of countries like India on addressing unilateral trade measures under the heading of 'response measures' at least provides an (admittedly small) window to start discussions on the design of BAMs under the UNFCCC.

The challenge of adding to an overburdened process also holds for discussing BAMs in the WTO, where the Doha negotiations already face major difficulties. It would be an 'extremely steep uphill battle' to reach agreement in the WTO, as consensus would be needed and this would include potentially affected countries.[279] Furthermore, members would need to overcome fears that discussions of BAMs would make their position in a possible dispute more fragile. If such an agreement were nevertheless possible, there are several options available to formalize it (meaning that this type of interaction management would become akin to treaty changes). Perhaps the least controversial measure would be to establish a committee on trade and climate change, in which members could notify other members of intended climate-related measures and discuss their legitimacy without resorting to dispute settlement.[280] Parties could also adopt a 'waiver' for certain types of BAMs.[281] Gary Hufbauer and colleagues go one step further, and suggest that climate-related trade measures could possibly be disciplined through the adoption of a plurilateral WTO Code.[282] They argue that '[a]lthough such a Code would require consensus of all WTO members to be formally added to the WTO agreement, such action could be politically possible because it would not necessarily require that all members agree to the text or

---

276 Werksman and Houser 2008.
277 UNFCCC Decision 1/CP.13, para. 1(b)(i).
278 Zelli and van Asselt 2010, 91.
279 Cosbey 2009, 22.
280 Epps and Green 2010, 255.
281 Charnovitz 2003, 157; Stokke 2004, 351; Hufbauer *et al.* 2009, 97.
282 Hufbauer *et al.* 2009.

substance of the code'.[283] While a plurilateral approach may seem sensible compared to a multilateral process involving each WTO member or UNFCCC party, to remain acceptable to developing countries it is crucial that they are involved in such an agreement.[284]

Another option would be to address the issue outside of both the climate and trade regimes, for example in the G8+5, the G20, the Major Economies Forum or the OECD. Each of these forums has its own advantages and drawbacks. Whatever forum is chosen, it is more realistic to start the discussions in an informal setting, and refrain from very technical discussions about the implementation of BAMs in the early stages.[285] This could be done in several forums simultaneously, including in the UNFCCC and the WTO, where the agenda item of response measures and the CTE provide entry points.

Although such discussions may not directly clarify the legality and environmental and economic effectiveness of BAMs, even 'an informal guideline might be very effective as it could enhance the WTO acceptability and certainty of any approach [and] could also be effective in limiting the use of border adjustments'.[286] In other words, although an international agreement may be preferable, informal discussions could raise awareness of the interactions and their potential consequences even in its absence. A starting point could be the guidance drafted by a group of experts on BAMs, which provides general guidelines with respect to their design.[287]

6. AUTONOMOUS INTERACTION MANAGEMENT

If international cooperation on border adjustment measures proves to be too controversial, autonomous interaction management is possible through the domestic design of BAMs. The design could address the concerns discussed in this chapter in various ways.

First, a transparent, extensive examination of the (sub-)sectors at risk of carbon leakage at a high level of disaggregation before BAMs are put in place could ensure that the measures are targeted at those (sub-)sectors for which leakage is a real and not just a perceived concern. For instance, while there is a solid research base providing an indication of the sectors

---

283 Hufbauer *et al.* 2009, 98.
284 Eckersley 2010, 387.
285 Neuhoff *et al.* 2008, 11.
286 Neuhoff *et al.* 2008, 11.
287 Cosbey *et al.* 2012.

at risk in the EU,[288] political factors influenced key choices determining the EU's decision to identify sectors at risk of carbon leakage.[289] Furthermore, as the various factors influencing carbon leakage (including the costs of climate policy and trade exposure) change over time, any list of sectors at risk would need to be dynamic.

Second, depending on the outcomes of such analyses and on sector- and country-specific characteristics, other options to address competitiveness and leakage, including lowering non-climate policy-related burdens for affected industries, enhancing the scope of a domestic emissions trading system and adopting government-led sectoral agreements, should not immediately be discarded.[290]

Third, administrative possibilities could be included in the design of the BAM to allow producers from other countries to provide evidence of the emissions from the production process and to appeal decisions that affect them directly.[291]

Finally, the timing of BAMs should provide ample time for governments to conclude an international climate change agreement that could make their use obsolete. Even proponents of BAMs will admit that the measures are only a second-best solution in the absence of a comprehensive international climate agreement. However, shortening the time frame of their adoption might completely preclude the possibility of reaching such an agreement in the first place.

7. CONCLUSION

The relationship between the climate and trade regimes continues to fascinate scholars and practitioners, irrespective of the fact that there is a 'pattern of accommodation between the regimes in ways that have prevented the outbreak of any direct collision of rules (or any conflicts involving the implementation of the rules)'.[292] This chapter has identified numerous areas of overlap, and has illustrated how this overlap leads to several potential policy (and possibly normative) conflicts and potential synergies. Some of the potential conflicts, such as limitations on the use of flexibility mechanisms, may emanate directly from the climate treaties, although neither the UNFCCC nor the Kyoto Protocol mandates

---

288 E.g., Hourcade *et al.* 2008; Reinaud 2008; De Bruyn *et al.* 2013.
289 van Asselt 2009, 66–69.
290 van Asselt and Biermann 2007.
291 Cosbey *et al.* 2012.
292 Eckersley 2009, 17.

specific trade measures. Most potential conflicts are due to the silence in the climate treaties on the use of unilateral trade measures accompanying domestic climate policies, combined with the unspecified nature of mitigation commitments under the Kyoto Protocol and a trend towards even more obscure bottom-up commitments. Whether these tensions between trade-related climate policies and WTO law will actually result in a normative conflict remains doubtful, however, and will depend both on how measures are designed and how specific WTO agreements would be interpreted by the WTO's dispute settlement bodies.

The chapter has shown how the issue of border adjustment measures may become the proverbial straw that breaks the camel's back, especially if climate change mitigation efforts are stepped up in a world of unequal carbon prices. BAMs are not, as a matter of principle, ruled out by the pertinent rules of international trade law. Past case law even suggests that such measures would stand a chance of being found admissible in a trade dispute; and in the event that a violation of free trade disciplines is found, the measure could be justified under the general exceptions of the GATT. However, uncertainties about the legality of border adjustment measures will persist and depend on their specific design.

At first blush, interaction management through legal techniques offers various tools to address the interactions between the environmental and non-environmental regimes. More specifically, the chapter underlines how in case of a dispute on BAMs before the WTO, various provisions in the GATT could be interpreted in light of the climate treaties. While various legal scholars have highlighted the potential of Article 31.3(c) of the VCLT and its 'principle of systemic integration' to harmonize different treaties in this regard, its theoretical attractiveness needs to be juxtaposed with the WTO's judicial practice so far, which shows only a limited willingness to take into account non-WTO law, including international environmental law. With respect to conflict resolution techniques, while a potential multilaterally agreed trade measure under the climate treaties could arguably be viewed as *lex specialis* vis-à-vis the WTO agreements, this ultimately depends on one's systematization of international law.

Interaction management through institutional coordination between the climate and trade regimes has been almost non-existent, perhaps reflecting their respective natures as an environmental and an economic regime. Nevertheless, coordination on specific overlapping issues may resolve some of the tensions, as this chapter has shown with respect to BAMs. Involving potentially affected countries in these discussions could enhance the legitimacy of climate-related trade measures and could enhance compatibility with both international trade and climate change law. In particular, combining the 'stick' of trade measures with 'carrots' for developing countries could make them more palatable. For BAMs, various questions regarding possible international coordination still need to be resolved, including the forum in which to discuss them (UNFCCC, WTO, an alternative forum or various forums at the same time), and the level of formality of any output (from a legally binding agreement to informal guidelines). In the absence of an international agreement on disciplining BAMs, I have listed various options through which countries intending to adopt BAMs can address the concerns raised. This discussion offers a reminder of the potential for autonomous interaction management.

PART IV

Synthesis and conclusions

9. Regime interactions in global climate governance

1. INTRODUCTION

This chapter takes a bird's eye view of regime interactions in global climate governance by discussing and contrasting the case studies and situating the observations and findings of the previous chapters in a broader context. The chapter reviews the different types of interactions that emerged from the case studies, as well as the possibilities and limitations of various ways of interaction management. The overall purpose of this chapter is to provide insights into the promises and pitfalls of legal techniques and institutional coordination in managing the outcomes of regime interactions in global climate governance.

2. TYPES OF REGIME INTERACTIONS

2.1 Objects of Regime Interactions

Existing studies of regime interactions have focused primarily on negotiated, treaty-based regimes, most of which constitute 'hard law'. This observation is largely valid also for studies of the interactions between the UN climate regime and other regimes. For example, in addition to the case studies examined in this book, treaty-based regimes covered in the literature on interactions in global climate governance include: the Ramsar Convention on Wetlands;[1] the Montreal Protocol on ozone depletion;[2] the World Heritage Convention;[3] international investment

[1] E.g., Pittock 2011.
[2] E.g., McCabe 2007; Oberthür *et al.* 2011.
[3] E.g., Burns 2009.

treaties;[4] human rights instruments;[5] international law on refugees;[6] and the law of the sea.[7]

This book in part echoes the dominance of this hard law focus through its focus on interactions involving the biodiversity and trade regimes. And yet there are also important and complex interactions between soft and hard law in global climate governance. If the focus is squarely on treaty-based regimes, these interactions (and the associated consequences) could easily become a blind spot in analyses of regime interactions. The view that soft law matters in international governance should thus be extended to acknowledge the ways in which soft law interacts with hard law. The case studies support this argument in two distinct ways.

First, the case of minilateral clean technology agreements confirms the view that soft law and hard law can interact with each other in ways that have so far have received insufficient scrutiny. As noted in Chapter 3, studies have started to shed light on how soft law and hard law may interact beyond the models initially developed by international lawyers (i.e., soft law developing into hard law over time; or occupying the niches left by hard law) and have indicated that soft law may also act as an antagonist to hard law.[8] Chapter 6 showed that similar observations can be made with respect to minilateral clean technology agreements with a soft law nature. In the area of global climate governance, we are witnessing 'a proliferation of formal and informal lawmaking forums, each championed by a diverse coalition of States and non-State actors, and each competing to formulate international rules and norms'.[9] While a 'focal institution'[10] already existed (i.e., the UN climate regime), minilateral initiatives such as the APP were driven by powerful actor coalitions that pursued an approach to international climate policy that diverged from the Kyoto Protocol. The US, in particular, has played an instrumental role in various initiatives since 2005. In the case of the APP this had direct and indirect consequences for the UN climate regime. While the Kyoto Protocol and minilateral clean technology agreements can be seen as competitors to the extent that their core approaches differ (with the former adopting mandatory targets and timetables, and the latter

4 E.g., Werksman *et al.* 2003.
5 E.g., S. Humphreys 2009; Cameron and Limon 2012.
6 E.g., Williams 2008; McAdam 2012.
7 E.g., K. Scott 2005.
8 Shaffer and Pollack 2010; 2011.
9 Pollack and Shaffer 2012, 250.
10 Jupille and Snidal 2006, 12.

adopting a voluntary, technology-oriented approach), in other ways (e.g., data collection, knowledge exchange, and capacity building) they can clearly complement each other. Chapter 6 thus illustrated that the relationship between soft law and hard law in global climate governance is not easily classified as either complementary or antagonistic and that different types of interaction between soft and hard law can take place simultaneously.

Second, and perhaps more controversially, soft law–hard law interactions may also take place in the context of what positivists would consider pure hard law interactions. While the interactions between the CBD and the Kyoto Protocol could be seen as an exemplary case of hard law interactions, with both undeniably being treaties in the sense of the VCLT, the nature of the obligations contained in the treaties is quite different. As Stuart Harrop notes, the CBD is a 'purportedly normative hard law that operates in the manner of an aspirational, policy-oriented soft instrument'.[11] In part, this is to be expected from an environmental framework convention, and the UNFCCC itself is not a shining example either of an agreement with unambiguous and precise obligations backed up by a strong compliance mechanism. However, whereas the CBD has largely failed to 'harden' its obligations over the years,[12] parties in the climate regime managed to develop legally binding, concrete commitments through the Kyoto Protocol (particularly for the developed countries) and introduced a compliance mechanism (although not one backed up by strong sanctions). To use the terminology of Kenneth Abbott and Duncan Snidal,[13] the climate regime has hardened over time in terms of its level of (legal) obligation, precision, and delegation.[14] The forest-related interactions examined in Chapter 7 can be related to these differences in 'hardness': the Kyoto targets (exerting a relatively high level of legal obligation) and the (relatively precise) modalities on forest carbon sinks provided incentives for behaviour that could be inconsistent with the objectives of the biodiversity regime, while the biodiversity regime did not provide countervailing incentives because of its lack of obligation and precision. Although the CBD work programme for forest biodiversity has gradually become more concrete, and the Aichi targets (which include targets relevant for forest biodiversity) seem to signal an intention to harden the CBD, the targets remain largely aspirational.[15] So

11 Harrop 2011, 119; see also Harrop and Pritchard 2011, 475–476.
12 Harrop and Pritchard 2011, 476–478.
13 Abbott and Snidal 2000.
14 von Stein 2008, 246–248.
15 Harrop and Pritchard 2011, 479.

while the biodiversity regime is undeniably treaty based and as such could be seen as hard law, the key point here is that interactions between two hard law instruments (in this case the CBD and the Kyoto Protocol) may sometimes have more in common with soft law–hard law interactions than with interactions between hard law instruments only, reinforcing the case for extending the study of regime interactions to soft law.

2.2 Causal Mechanisms of Regime Interactions

By distinguishing between different causal pathways for interactions, Sebastian Oberthür and Thomas Gehring have provided a useful foundation for identifying how different regimes influence each other (see Chapter 4).[16] With the exception of impact-level interactions, whose existence is empirically difficult to verify, the various causal mechanisms of interactions they identify can be observed in the three case studies in global climate governance.

The mechanism of *cognitive interaction* can be observed in the case of interactions between minilateral clean technology agreements and the UN climate regime. More specifically, the APP has acted as a 'policy model', in which 'members of the target institution [i.e., the UN climate regime] voluntarily use some aspect of the source institution [i.e., the APP]'.[17] Chapter 6 highlighted several features of minilateral clean technology agreements that were used voluntarily by countries like Australia, Japan and the US in the climate change negotiations. Some of these features have found their way into the UN climate regime, such as the approach based on (voluntary) pledges that was adopted in the Copenhagen Accord and consolidated a year later in the Cancún Agreements, and which may well find its way into a 2015 agreement (see Chapter 2). This does not mean that the minilateral clean technology agreements in which Australia, Japan and the US were participating are the sole source of the pledge-and-review approach adopted in Copenhagen and Cancún, as it is difficult to discern any clear evidence that would point to such a causal relationship. However, the minilateral initiatives reinforced the negotiation position adopted by these countries in the UNFCCC context.[18] The consequences of this cognitive interaction are still unclear. Although the 'policy model' type of interaction is generally expected to strengthen the effectiveness of the target regime,[19] it remains to be seen whether the

16 Oberthür and Gehring 2006b.
17 Cf. Gehring and Oberthür 2006a, 328.
18 McGee and Taplin 2009b; Vihma 2009.
19 Cf. Gehring and Oberthür 2006a, 329.

model of minilateral clean technology agreements will enhance the effectiveness of the UN climate regime.

A clear instance of *interaction through commitment* can be identified in the WTO case study.[20] The fact that neither the UNFCCC nor the Kyoto Protocol (unlike, for instance, the Montreal Protocol or the Convention on International Trade in Endangered Species) contain or condone climate-related trade measures but rather seem to defer to the WTO can be seen as an instance of the 'chilling effect'.[21] This interaction through commitment emerged because the two regimes cover two different policy domains (liberalizing trade versus reducing climate risks), while their memberships overlap to a large extent. What followed was an attempt at 'jurisdictional delimitation'[22] in the form of provisions in the climate treaties calling for the minimization of trade impacts. These provisions do not clarify which regime has regulatory authority in the case of climate-related trade measures, but at least create the appearance (for now) that the two regimes coexist harmoniously.[23]

The mechanism of *behavioural interaction* features in all three case studies. First, the disincentives provided by the creation of the APP for participation in the climate treaties (especially the Kyoto Protocol) and the inconsistencies between the APP and the climate treaties in terms of differentiation, impacts and adaptation, and stakeholder participation, are two examples of behavioural interaction that led to a potential policy conflict between the two regimes. However, as Chapter 6 showed, another behavioural interaction resulted in a positive outcome in that the actions by states and non-state actors participating in the minilateral clean technology agreement indirectly supported the implementation of the climate treaties.

Second, the interactions between the climate and biodiversity regimes are examples of behavioural interaction, but in these cases the UN climate regime is the source of the interactions rather than the target. As Chapter 7 shows, the interactions originated in the decision-making process of the climate regime, first with respect to the inclusion of forest carbon sinks in the CDM and later with respect to REDD. In the case of forest carbon sinks in the CDM, the decisions provided incentives for parties to implement the Kyoto Protocol in a way that could hamper the effective implementation of the CBD. With regard to REDD, the rule development has not yet reached a stage where such a conclusion can be

---

20 Cf. Oberthür 2006, 57.
21 Zelli and van Asselt 2010, 88; Gehring 2011, 236.
22 Gehring and Oberthür 2006a, 335–339.
23 Cf. Eckersley 2009.

drawn, but in this case too the consequences of interactions will depend on the incentives included in the decisions adopted by the parties to the UNFCCC and the subsequent behavioural changes by parties.

Finally, by affecting the extent to which individual parties to the climate treaties resort to climate-related trade measures, the WTO indirectly influences the development and performance of the UN climate regime. Chapter 8 highlighted this behavioural interaction for several possible trade-related policies and measures that could be adopted by parties to the UNFCCC and/or the Kyoto Protocol. The influence is perhaps most visible in the design of border adjustment measures. While no such proposal was adopted by either the EU or the US, the detailed legislation proposed in the US was drafted with an eye on compatibility with WTO law and practice, and the potential violation of trade law disciplines is often used as a justification not to adopt them.

2.3 Intentionality of Regime Interactions

By introducing the notion of 'political linkages', Oran Young drew attention to the fact that not all interactions appear out of thin air.[24] Indeed, as Chapter 4 noted, regime interactions can be either intentional or unintentional.[25] However, this dichotomy is hard to verify empirically, since it can be difficult to identify the actors driving interactions, let alone (objectively) their intentions. Nevertheless, the cases studies provide some insights into the intentionality of regime interactions in global climate governance.

The existence of intentionality is clearest in the case of minilateral clean technology agreements. Although it may be challenging to identify the precise intentions of the proponents of minilateral initiatives, there are clear indications that the APP was initiated by the US and Australia to influence the UN climate regime. These indications include its timing (a few months after the Kyoto Protocol entered into force); its participants (including the two most notable Kyoto non-parties at the time); and its key features (in line with the climate policy proposals of these countries).[26] As I have argued with colleagues, the APP and similar initiatives 'were created not out of ignorance of the climate regime but *because* of it'.[27]

---

24 O.R. Young 2002, 23.
25 Zelli 2009.
26 Karlsson-Vinkhuyzen and van Asselt 2009, 201–202.
27 Biermann *et al.* 2009, 24.

The APP can be regarded as an effort 'on the part of major actors to exploit interactive decision making to promote their own ends regardless of the consequences in terms of the common problem'.[28] The countries initiating the APP made use of opportunities for forum shopping,[29] with a view to causing a 'strategic inconsistency'[30] with the climate regime. The precise 'consequences in terms of the common problem' remain unclear, because it is too early to rule out the possibility that the approach adopted by minilateral clean technology agreements like the APP will be less effective in addressing climate change than the Kyoto model. Still, Chapter 6 showed that the actions of the US and Australia led to (potential) policy conflicts with the climate regime. This conflictive outcome of the intentional interaction may seem surprising. As Gehring and Oberthür note, 'intentionally triggered cases of disruption appear to be particularly rare' in global environmental governance. However, they add that 'disruptive interaction is occasionally employed intentionally even in environmental governance to bring about change within other institutions'.[31] This observation seems to hold true in the case of the interactions between minilateral clean technology agreements and the UN climate regime.

The intentionality of regime interactions between the UNFCCC and the CBD or the WTO is much harder to establish in the absence of detailed studies of the motivations of various actors and actor coalitions to adopt the decisions that triggered the interactions. With respect to the interactions between the CBD and the climate regime, it is likely that the effects were largely unintentional.[32] While the decisions on forest carbon sinks adopted by the treaty bodies of the climate regime may have failed to provide adequate biodiversity protection, they do not provide evidence of any of the Kyoto Protocol parties' intentions to undermine the objectives of the CBD. On the contrary, the mere existence of safeguards points to awareness that conflicts needed to be avoided, even though the incentives provided by these safeguards turned out to be quite weak. With respect to the interactions related to REDD, awareness of the potential interactions is again present, as suggested by the biodiversity-related safeguards in the UNFCCC decisions. Again, although it is difficult to prove the intentions of the parties to the climate regime, the inclusion of biodiversity safeguards (albeit very general ones) could be viewed as

---

28 O.R. Young 2002, 133.
29 Alter and Meunier 2009, 16; see also Helfer 2004.
30 Raustiala and Victor 2004, 298.
31 Gehring and Oberthür 2006b, 13.
32 Cf. Oberthür 2001, 7.

evidence of parties' intentions to capture synergies with the CBD, rather than an intention to instigate conflicts.

With respect to the WTO, it is safe to assume that 'international trade rules were not adopted in order to slow down or "chill" the development of effective international climate policy',[33] or international environmental policy more broadly. Similarly, the climate treaties were not adopted to interfere with international trade policy, as affirmed by its provisions related to trade. It can be argued, however, that this approach of non-interference with trade policies was consciously adopted by the countries negotiating the UNFCCC.[34] Furthermore, if BAMs were adopted, this could more clearly be identified as an intentionally triggered interaction.

3. CONSEQUENCES OF REGIME INTERACTIONS

3.1 Conflicts

In Chapter 4, I introduced the distinction between normative and policy conflicts. The former refers to a conflict of norms in the strict legal sense. Several definitions have been put forward by international lawyers, showing that it is not settled when such a conflict can be established. Still, even when applying one of the broadest constructions of 'normative conflict', the definition proposed by Erich Vranes that '[t]here is a conflict between norms, one of which may be permissive, if in obeying or applying one norm, the other norm is necessarily or potentially violated',[35] it is notable that no such conflicts could be identified in the preceding chapters, arguably affirming the 'presumption against conflict'.

For minilateral clean technology agreements, the conclusion that there is no actual or potential normative conflict can be drawn simply because one of the interacting regimes is not treaty based. The definitions proposed by international lawyers all have in common that they apply only to conflicts between norms from treaty-based regimes.

In the same vein, it is not possible to establish a normative conflict between the climate treaties and the CBD. Although Concetta Maria Pontecorvo refers to the interaction between the CBD and the UN climate regime related to the inclusion of forest carbon sinks as a

33 Oberthür 2001, 13.
34 Eckersley 2009, 17.
35 Vranes 2009, 38.

'potential "norm conflict"',[36] this assertion is questionable even under the broad legal definitions. The climate treaties and the CBD do not contain inconsistent or incompatible norms on the use of forest carbon sinks; they do not contain any specific norms on this issue at all.[37] Chapter 7 shows that the regime interactions rather stem from the decisions adopted under the climate treaties. The existence of a normative conflict is thus conditional on viewing such treaty body decisions as an integral part of the treaties, a matter that remains to be clarified (see below). But even if one accepted this point, it is doubtful that a normative conflict would come into existence. The COP decisions related to forest carbon sinks do not contain unconditional permissions that would necessarily or potentially violate the provisions of the CBD. Moreover, it is challenging to identify which norms of the CBD, which consists of 'a pastiche of vague commitments, ambiguous phrases, and some awkward compromises',[38] would be violated.

In the case of interactions between the climate and trade regimes it is also not possible to identify a normative conflict.[39] The reason for this is that the climate treaties do not contain norms prescribing or permitting specific conduct that would violate WTO law, but rather list a (non-exhaustive) range of policies and measures that states can adopt to achieve the treaties' objectives. This does not mean that normative conflicts can be ruled out completely. First, as discussed in Chapter 8, a 'horizontal'[40] normative conflict (a direct conflict between a climate treaty and a trade agreement) could arise if parties adopt a future climate agreement that permits or requires trade measures, for instance, against non-compliers or non-parties.[41] Second, unilateral trade measures that seek to implement the climate treaties could still lead to 'vertical' conflicts with WTO law.[42] Chapter 8 highlighted how the design of unilateral trade measures could affect (and increase) the chances of such a vertical conflict. However, beyond the fact that no such measure has been adopted yet, it also remains uncertain whether their adoption would actually violate WTO law, as the judgment of the WTO dispute settlement bodies would depend on the measures' design. Focusing on a different example (the adoption of various technical measures in EU

---

36 Pontecorvo 1999, 731.
37 van Asselt *et al.* 2008, 430.
38 Raustiala and Victor 1996, 19.
39 van Asselt *et al.* 2008, 434.
40 Vranes 2009, 374.
41 van Asselt *et al.* 2008, 434.
42 Vranes 2009, 374.

climate law), Vranes also concludes that 'there are no conflicts of norms in the technical sense between WTO obligations and the Kyoto Protocol's prescriptions on domestic climate mitigation measures'.[43] In short, while there is potential for normative conflict between the climate treaties and the WTO, there is no such conflict at the moment.

Moving away from the more narrowly defined normative conflicts, it is possible to identify several actual or potential policy conflicts in the three case studies, following the indicators for conflict outlined in Chapter 4.

In terms of *actual* policy conflicts, Chapter 7 highlighted the existing tensions between the implementation of the Kyoto Protocol's flexibility mechanisms and the CBD. The finding of such a policy conflict in itself is nothing new, but it does raise a new question of when a policy conflict comes into existence. Pontecorvo,[44] as well as Rüdiger Wolfrum and Nele Matz,[45] refer to the conflict as 'potential' on the basis that such a conflict will only materialize if actors (governmental or non-governmental) implement climate mitigation measures that have (visible) adverse effects on biodiversity. Establishing whether this is the case poses novel empirical challenges. It would require corroboration, for instance, that implementing specific CDM projects would result in negative impacts on biodiversity. While examples of such projects exist,[46] the next question is whether any adverse impact on any species or ecosystem would already be sufficient to establish a policy conflict between two regimes (i.e., when is a threshold crossed?). Rather than following such a line of inquiry, I would argue that the policy conflict *already* exists simply because the rules agreed upon by the Kyoto parties provide economic incentives that *could* (but not necessarily *would*) lead to negative biodiversity impacts and thus undermine the effective implementation of the CBD.

In line with Oberthür, who observed that interactions between the climate regime and non-environmental institutions resulted more frequently in conflicts than the interactions with environmental institutions,[47] the actual policy conflict between the climate and trade regimes can be traced back to their diverging objectives. Although these objectives are not necessarily conflicting, trade liberalization and climate protection generally pull in different directions. Moreover, the trade regime is usually considered as the 'stronger' regime, particularly

---

43 Vranes 2009, 398.
44 Pontecorvo 1999, 731.
45 Wolfrum and Matz 2003, 82.
46 Bäckstrand and Lövbrand 2006, 65.
47 Oberthür 2006, 59.

because of its dispute settlement system.[48] The provisions in the climate treaties demanding that measures adopted in pursuit of climate goals are compatible with international trade law and minimize the impacts on international trade can be seen as signs of the 'chilling effect' in practice. Again, while it is difficult to empirically prove that an absence of climate-related trade measures (adopted either multilaterally or unilaterally) has resulted in reduced climate change mitigation action, for establishing the actual policy conflict I would argue that it is sufficient that the objectives diverge.[49]

In terms of *potential* policy conflicts, the case studies pointed out several tensions below the surface. Chapter 6 noted several potential conflicts between minilateral clean technology agreements and the climate regime. The first of these is only a potential conflict to the extent that it could be said that actors in the climate regime learn lessons from such agreements that reduce the effectiveness of the climate regime.[50] This potential conflict is rather hypothetical, however, because it is difficult to establish whether minilateral clean technology agreements would indeed form a 'policy model' for the climate regime or whether it boils down to the same actors pursuing the same strategies in different regimes. Moreover, it is hard to ascertain whether the adoption of key features of the minilateral clean technology agreements (e.g., the number of parties or the technology-oriented approach adopted) would undermine the objectives of the UN climate regime or rather support them.

The potential outcomes of two other interactions between minilateral clean technology agreements and the UN climate regime could more clearly be seen as conflictive. As explained in Chapter 6, the APP reduced the attractiveness of participating in the Kyoto Protocol, thereby diminishing the effectiveness of the climate regime. The other potential policy conflict was related to the divergences between the APP and the climate treaties in terms of principles and concepts. Notably, the lack of differentiation between developed and developing countries formed a clear departure from the UNFCCC's principle of common but differentiated responsibilities.

Chapter 7 showed that the potential policy conflict between the climate regime and the CBD concerning the implementation of REDD depends on the design of the mechanism. This refers not only to the biodiversity safeguards that have so far emerged in UNFCCC decisions, but also to

---

48 Kulovesi 2011b, 59.
49 Cf. Oberthür 2001, 13.
50 Cf. Gehring and Oberthür 2006a, 329.

the funding modalities that still need to be clarified. If climate finance for REDD were linked to reduced carbon emissions only, something that its very name seems to imply, the risks for biodiversity will likely be higher.

Table 9.1 sums up the different normative and policy conflicts identified in the three case studies, with reference to the specific indicators for conflicts proposed in Chapter 4.

Table 9.1 Types of conflicts involving the UN climate regime

	Normative conflict	Policy conflict
Actual	No conflicts identified	*CBD: Opposing economic incentives* Economic incentives of Kyoto's flexibility mechanisms hampered effective implementation of the CBD *WTO*: *Diverging objectives* Incompatibility of trade liberalization and climate protection objectives led to 'chilling effect'
Potential	*WTO: Incompatible norms* Trade measures in future climate treaty could violate WTO law *WTO: Incompatible norms*: Unilateral trade measures could violate WTO law	(*Minilateral clean technology agreements: Negative diffusion and learning* To the extent lessons are learned that reduce the effectiveness of the future climate regime) *Minilateral clean technology agreements: Opposing economic incentives* APP could have made it more attractive to not participate in Kyoto Protocol or future climate treaty *Minilateral clean technology agreements: Different principles and concepts* APP's exclusion of differentiation; adaptation and impacts; and stakeholder participation could lead to inconsistencies with UNFCCC *CBD: Opposing economic incentives* REDD finance targeted only at carbon emissions could hamper effective implementation of the CBD

3.2 Synergies

In addition to actual or potential conflicts, the previous chapters have showed that regime interactions in global climate governance have also led to synergies.

Initial studies on the relationship between minilateral clean technology agreements and the climate regime focused on the negative impacts.[51] Yet minilateral clean technology agreements may also contribute to the effective implementation of the multilateral climate regime. This should not be surprising, as the objectives of several minilateral clean technology agreements overlap to a large extent with those of the climate regime, even though their approaches to climate change mitigation diverge. The APP contributed to the climate regime's objectives by collecting sectoral data in the partner countries, capacity building in, and knowledge sharing with, the developing partner countries and helping developed countries to implement their commitments, especially those in the area of technology development and transfer.

The actual synergies between the CBD and the climate regime primarily stem from their overlapping *ratione materiae*: both regimes are concerned with the conservation of ecosystems. For the CBD, this follows directly from its stated objective. For the UNFCCC and the Kyoto Protocol, this goal can be indirectly derived from various provisions in both treaties.

Whether the inter-institutional learning between minilateral clean technology agreements and the climate regime results in synergies is, as discussed above, dependent on whether such learning can be observed in the first place and on participants in the UN climate regime learning lessons that will further its performance. This could be the case, for instance, if it turned out that a minilateral approach involving only the world's major emitters were indeed to result in faster and earlier emission reductions, as proponents of the APP have argued. However, any finding of such a synergy is speculative.

Given their joint origin in the 1992 Earth Summit and their overlapping objectives and concepts, it is perhaps not surprising that there are various potential synergies between the CBD and the climate regime. One of these is rooted in the possibility of providing common incentives for reducing deforestation. It is clear that reducing deforestation as such could result in both biodiversity and climate benefits. These benefits could be even greater if REDD finance not only financially rewards emission reductions, but also contributes to biodiversity protection. Another potential synergy is related to the opportunities for enhancing the efficiency of both the climate and biodiversity regimes through common monitoring and reporting obligations that are relevant to both regimes. Although monitoring and reporting of the biodiversity impacts

[51] E.g., McGee and Taplin 2006; Lawrence 2007; van Asselt 2007.

of REDD is not yet streamlined, the CBD secretariat has highlighted this potential (see Chapter 7).

Finally, in the overview of interactions between the climate and trade regimes, I noted several potential synergies. More specifically, cuts in subsidies for fossil fuels could reduce unfair advantages for domestic producers while holding large potential in terms of emission reductions. Furthermore, the ongoing trade negotiations aim, among other things, to remove barriers to the trade in environmental goods and services including climate-friendly ones. While there is currently no consensus about which goods and services can be considered climate friendly, liberalization of trade in such goods and services could also serve the goals of both regimes.

Table 9.2 provides an overview of the synergies identified in the three case studies, with reference to the specific indicators for synergies introduced in Chapter 4.

Table 9.2 Types of synergies involving the climate regime

	Synergies with the climate regime
Actual	*Minilateral clean technology agreements: Shared supporting measures* APP contributed to implementation of climate treaties through data collection, knowledge exchange and capacity building
	CBD: Shared principles and concepts Both regimes are (directly or indirectly) concerned with, and contribute to, the conservation of ecosystems
Potential	(*Minilateral clean technology agreements: Positive diffusion and learning* To the extent lessons are learned that enhance the effectiveness of the climate regime)
	CBD: Common economic incentives Finance targeted at both carbon emissions and biodiversity impacts could enhance biodiversity protection
	CBD: Streamlined monitoring and reporting Common monitoring and reporting guidelines for the biodiversity impacts of climate measures by the CBD could support the UNFCCC
	WTO: Common economic incentives Reducing fossil fuel subsidies and removing trade barriers for climate-friendly goods and services under the WTO could lead to climate change mitigation benefits

4. MANAGEMENT OF INTERACTIONS

Having outlined the different types of regime interactions in global climate governance emerging from the three case studies as well as their consequences, this section examines the ways in which these interactions have been managed (or could be managed), with a view to identifying the potential and limitations of different ways of interaction management.[52]

4.1 Legal Techniques: Opportunities

Harmonious treaty interpretation

'Prevention is better than cure', the saying goes. And indeed, in addressing regime conflicts, there is much to be said for legal techniques that seek to avoid normative conflict. As Chapter 5 discussed, the notion of systemic integration contained in Article 31.3(c) of the VCLT, instructing interpreters to take into account 'any relevant rules of international law applicable in the relations between the parties', has received particular attention as a means of harmonious treaty interpretation. In this way, normative conflicts could be avoided entirely,[53] and policy conflicts could be mitigated. The previous chapters confirm that the principle holds potential to address the conflictive outcomes of regime interactions.

Chapter 7 showed that the principle could be used to avoid conflicts in international environmental law by affirming a duty for states in the implementation stage to interpret the Kyoto Protocol in a way that makes it compatible with commitments in the CBD.[54] On a basic level, the principle thus serves as a reminder to states that they have entered into several commitments, which all need to be respected simultaneously. The texts of the climate and biodiversity treaties are sufficiently ambiguous to allow states to fulfil their various obligations at the same time.

Chapter 8 further highlighted the potential of applying the principle in dispute settlement by explaining how the climate treaties could also be of use in the interpretation of other ambiguous or indeterminate WTO norms in a climate–trade dispute before the WTO.[55] A country adopting the measure could invoke Article 31.3(c), and use its ratification of climate treaties in its defence of the non-commercial, environmental objectives of its measure. All other things being equal, participation in

[52] This section draws on van Asselt 2012a.

[53] Aust 2007, 230; see also Wolfrum and Matz 2003, 133–147.

[54] Pontecorvo 1999, 741.

[55] van Asselt *et al.* 2008, 435–436.

climate treaties would make the tests formulated in the exceptions of the GATT easier to meet.

It remains unclear, however, whether Article 31.3(c) is a 'master-key'[56] for dealing with conflictive regime interactions. Chapter 8 showed that other treaty interpretation devices, such as the object and purpose of treaties or their ordinary meaning, may also allow for the balancing approach that is needed for harmonious treaty interpretation. From a practical perspective, this argument was underscored by the fact that WTO dispute settlement bodies have been very reluctant to seek recourse to Article 31.3(c) and, in the rare cases when they have done so, have used it in a restrictive fashion, as exemplified by the requirement from the *EC-Biotech* case that two treaties need to have parallel membership. However, the upshot of assigning only 'residual value'[57] to Article 31.3(c) has been a rather arbitrary inclusion of primarily non-WTO law that seems to prioritize market liberalization over other policy objectives.[58]

An important observation is that treaty interpretation is not confined to international adjudicators, even though it is often discussed in this context. On a daily basis, government officials implementing specific agreements will need to interpret treaties and may do so in a way that aims at harmonizing seemingly diverging obligations, within the limits of the VCLT.[59] Even if one accepts the requirement of parallel membership, this 'limitation applicable in judicial contexts has no relevance when states parties to a treaty themselves interpret their rights and obligations under that treaty'.[60] States are thus free, and could be said to have a duty, to ensure that treaties are interpreted taking into account their wider normative environment.

For international adjudicators, pursuing harmonious treaty interpretation is more contentious, because dispute settlement bodies 'explicitly state the degree to which and the foundations upon which norms from different regimes have been taken into account', which may have implications for 'the further development of a more or less systemic vision of public international law as a whole'.[61] This is one of the reasons why the WTO dispute settlement bodies have been so reluctant to use the

56 McLachlan 2005, 280–281.
57 Van Damme 2009, 375; Samson 2011, 711–712.
58 Dunoff 1998, 24–25; Kulovesi 2011b, 167–170.
59 Matz-Lück 2012, 214.
60 Samson 2011, 710.
61 Matz-Lück 2012, 207–208.

potentially strong integrative device contained in Article 31.3(c). Therefore, the consequences and risks of proposals to disregard the requirement of parallel membership altogether by only requiring that parties involved in a dispute are members of both regimes should not be understated.[62] Specifically, if parties to a dispute differ, interpretations of the same provision may diverge, and extraneous rules may start to become influential even for those states that have not consented to them. Such divergences could, however, be reduced by clarifying the basis on which extraneous rules are selected. In this regard, the ILC follows Pauwelyn's suggestion, stating that

> it might also be useful to take into account the extent to which that other treaty relied upon can be said to have been 'implicitly' accepted or at least tolerated by the other parties 'in the sense that it can reasonably be considered to express the common intentions or understanding of all members as to the meaning of the … term concerned'.[63]

Following such an approach could increase the usefulness of Article 31.3(c) by expanding the number of possible extraneous rules that could be taken into account in the interpretation process undertaken by dispute settlement bodies. At the same time, it highlights that not every exogenous rule applicable between the parties should be given (equal) interpretative force. Scholars like Margaret Young and Robert Howse have stressed the importance of identifying *why* some exogenous rules are more legitimate than others.[64] Among the factors that could determine this, Young includes accessibility for the wider WTO membership, the breadth of support for an international organization, its constituency (including developed and developing countries), as well as the transparency and openness of its decision-making process.[65] Her suggestion moves away from the narrow construction put forth in *EC-Biotech* and opens the door for enhancing the usefulness of legal techniques in addressing overlaps involving the trade regime, including the interactions between the climate and trade regimes. It also underlines the importance of examining the origins of the norms of the climate regime, and the extent to which they are supported by both state and non-state actors.

---

[62] ILC 2006, para. 472.
[63] ILC 2006, para. 472, citing Pauwelyn 2003, 257–263.
[64] Howse 2007, 41–42; M.A. Young 2011b, 237.
[65] M.A. Young 2011b, 237.

From conflict to interaction clauses

The climate treaties contain several provisions that regulate their relationships with other multilateral agreements and international organizations. For instance, with respect to the Montreal Protocol, the UNFCCC and the Kyoto Protocol delimit their scope by only covering 'greenhouse gases not controlled by the Montreal Protocol'.[66] While this provision shows awareness of the linkages between the problems of (and solutions to) climate change and ozone depletion, it does not in itself prevent or resolve conflicts between them. Chapter 7 further highlighted the Kyoto Protocol's reference to 'relevant international environmental agreements' in Article 2.1(a)(ii), but also showed that its instructions are not specific enough to avoid policy conflicts.[67] Chapter 8 pointed to the provisions of the UNFCCC and Kyoto Protocol calling for mutual supportiveness with the trade regime (Article 3.5 UNFCCC; Article 2.3 Kyoto Protocol), but again their phrasing does not provide guidance on how to solve conflicts involving climate-related trade measures.[68]

Of the other regimes studied in the previous chapters, only the CBD contains a clearly identifiable conflict clause, one which could impose limitations on climate measures.[69] But once again, the wording used in this provision ('serious damage or threat to biological diversity') does not provide sufficient guidance to parties and would require further elaboration, for example through a COP decision.

As noted in Chapter 5, there are various difficulties with the use of conflict clauses: their wording is often unclear and open to diverging interpretations (e.g., what does 'serious damage or threat to biological diversity' entail in practice?); they are not dynamic enough to reflect new developments (e.g., changes in scientific insights into the causes and impacts of climate change); and it is not always clear when a treaty comes into existence.[70]

Nevertheless, from a legal perspective, they provide the primary means of addressing the relationship between treaties. Their drafting thus provides an opportunity for managing regime interactions *ex ante*. Whenever a new treaty or amendment is negotiated (either within the UNFCCC context or outside of it) conflict clauses could be drafted in a way that fully considers the implications for other treaties, and preferably

---

66 E.g., UNFCCC, Art. 4.1(a); Kyoto Protocol, Art. 5.1.
67 Jacquemont and Caparrós 2002, 178.
68 Voigt 2008a, 299.
69 Wolfrum and Matz 2003, 124.
70 Vierdag 1988.

in an unambiguous manner.[71] Hence, drafting a list of all international legal instruments that may have an impact on the treaty under negotiation is a sensible starting point. Under such a 'stop and think approach'[72] the impacts of a new treaty or a treaty amendment would have to be carefully assessed, where appropriate, in cooperation with the relevant states, secretariats and international organizations. This suggestion is certainly not new. In 1953, Wilfred Jenks already noted the importance of consultations before and during the drafting of legal instruments.[73] However, there is as yet no standard procedure for treaty negotiators to assess the impacts of a new instrument on existing ones, or to consider how an instrument could contribute to the objectives of other treaties. An opportunity thus lies in introducing such an 'impact assessment' procedure in drafting new climate-related agreements. The procedure would need to take into account the dynamic nature of international (environmental and climate) law, and acknowledge that regime interactions are not static. This could be done, for instance, by providing a clear mandate for an inter-institutional dialogue, which would bring us into the realm of institutional coordination.

Chapter 7 showed that there is ample scope for drafting rules that promote the objectives of both the climate and biodiversity regimes, especially in the area of REDD. Similarly, Chapter 8 drew attention to options for amending the WTO agreements with a view to accommodating climate change considerations. There is thus scope for *legal* techniques to do so. This does not mean that there is necessarily sufficient *political* will to draft such clauses. Nevertheless, political will could possibly be created because 'the positive effects of mutually supportive clauses should benefit both regimes, and thus … be in the interest of parties on either side'.[74] Unambiguously drafted 'interaction clauses' could expressly reflect the intention that the two environmental treaties support each other and could give a mandate to treaty bodies for how such mutual supportiveness might be better achieved.[75] Furthermore, a mandate for treaty bodies to continue to consider the linkages with other agreements could provide for a dynamic element, which is crucial in a field where scientific insights and political developments are ever changing and regimes need to adapt to changing circumstances.

71 Wolfrum and Matz 2003.
72 Hicks 1999, 1669.
73 Jenks 1953, 452.
74 van Asselt *et al.* 2008, 432.
75 van Asselt *et al.* 2008, 431.

4.2 Legal Techniques: Limitations

Informality in international lawmaking

The climate regime, like many other international environmental regimes, is characterized by a form of lawmaking that departs from the traditional idea of treaty-based lawmaking. Under modern international environmental regimes, lawmaking does not stop when the treaty text is agreed upon, but continues through the decision-making bodies constituted by those treaties (e.g., the UNFCCC COP).

The consequence of this novel form of international lawmaking is that the 'trigger' of a regime interaction could take the form of a treaty body decision rather than the treaty itself. As I showed in Chapter 7, the potential conflict between the climate and biodiversity regimes on forest carbon sinks originated in the decisions made by the climate COP.[76] Yet legal techniques are primarily concerned with conflicts stemming from treaties as such. This focus is in line with discussions on the subject of conflicts in international law, which have centred on treaties as the source of conflict. For instance, Jenks's classic definition states that a 'conflict in the strict sense of direct incompatibility arises only where a party to the two *treaties* cannot simultaneously comply with its obligations under *both treaties*'.[77] In addition, various legal techniques to avoid or resolve conflicts are based on, or linked to, the law of *treaties*. Later definitions of 'conflicts of norms' are broader but even they may be insufficient to cover the various kinds of incompatibilities that may appear in international environmental law. Consequently, the question of how to deal with cases where the texts of two treaties are perfectly compatible, but subsequent rule development under one of the treaties leads to a conflict, is largely ignored.

This observation necessitates an examination of the legal status of the decisions of MEA treaty bodies: to what extent do such decisions constitute international lawmaking in a traditional sense? As noted in Chapter 5, this question has been the subject of debate among international lawyers.[78] The main message emerging from this debate is that the legal status of COP decisions depends on the underlying treaty provisions upon which they are based, as well as on the wording of the decisions.[79] If there is an explicit authorization in the treaty, such

---

[76] van Asselt 2012b.

[77] Jenks 1953, 426 (emphases added).

[78] Churchill and Ulfstein 2000; Röben 2000; Brunnée 2002; Brunnée 2005; Wiersema 2009.

[79] Röben 2000, 373–374.

decisions could be considered to be legally binding. However, such authorizations are rare,[80] and cannot be found in the climate treaties. For instance, COP/MOP decisions on the details of the Kyoto Protocol's flexibility mechanisms, such as the use of forestry projects in the CDM, are not covered by an explicit authorization rendering them binding law.[81] Moreover, even if there were close linkages between COP decisions and underlying treaty provisions, this does not mean that the decisions themselves are subject to the law of treaties and the conflict resolution rules contained therein.[82]

Like interactions triggered by treaty body decisions, the legal techniques discussed in Chapter 5 are largely irrelevant in addressing soft–hard law interactions that involve non-treaty-based regimes, as there will not be a normative conflict in the narrow legal sense. In interpreting treaties, states are of course free to take into account any soft law agreements they have entered into,[83] but there is no legal obligation to do so.

The increasing relevance of treaty body decisions as well as soft law in international (environmental) lawmaking hence limits the usefulness of conflict resolution techniques. This does not mean that any interactions arising from such decisions cannot be dealt with, but rather points to the need to think about alternative means to manage them.

Normative hierarchy and synergies

While even the wide legal definitions of 'conflict' between norms are arguably insufficient in that they do not cover all types of conflict, it should be asked: insufficient for what purpose? In this regard, it is illustrative to cite one of Vranes' main objections to a narrow definition of normative conflict: 'The problematic consequence … is that conflicts maxims such as the *lex posterior* principle cannot come into play …'.[84] In other words, it is important to establish that there is a conflict if one wishes to apply techniques through which one of the norms would prevail. This argument is thus based on an assumption that one of the norms *should* prevail, or that the existence of conflicting norms in a particular situation is undesirable. This assumption can be driven by a need to provide legal certainty about which norms apply to a particular

---

80 See, e.g., Montreal Protocol, Art. 2.9(d).
81 Brunnée 2002, 24.
82 Wiersema 2009, 247.
83 Cf. Matz-Lück 2012, 217.
84 Vranes 2009, 19.

situation. In addition, the assumption that normative hierarchy is desirable can also be explained through the 'structural bias' of actors belonging to a specific regime. Put simply, from an environmental perspective, environmental norms should trump trade norms.[85] From the trade perspective, trade norms should naturally prevail.[86] These assumptions can be criticized on two important counts.

First, viewing inter-regime conflicts as conflicts of *norms* is not necessarily the best characterization. This is why Chapter 4 drew attention to broader 'policy conflicts'. It is helpful to consider the distinction made by Ralf Michaels and Joost Pauwelyn between conflicting norms and conflicting legal systems in this regard.[87] In the case of the former, norms are considered to be part of the same system, which makes it possible, in theory, to apply conflict resolution rules like *lex specialis* or *lex posterior*. However, inter-regime conflicts may actually have more in common with conflicts between different legal systems, for which 'it becomes difficult to devise a neutral perspective from which neutral conflict solutions could be formulated'.[88] There may be prima facie objective criteria for determining a hierarchy between regimes, such as the number of states party to a treaty or the nature of treaty obligations, but none of these criteria provide a foolproof foundation for a normative hierarchy.[89] For instance, a high number of state parties could be explained by the importance of the subject matter of a treaty, but could equally be explained by shallow commitments contained in such a treaty. Similarly, the real 'higher aims' of a treaty vis-à-vis other treaties are difficult to locate and to objectively verify.[90] In other words, there is arguably no objective way in which to determine which norm should prevail (see also Chapter 10).

Second, pursuing a normative hierarchy distracts from the idea that different treaties may pursue similar or overlapping objectives and may actually lead to mutually supportive outcomes. This observation is especially pertinent in the case of conflict between two regimes within the same policy domain, such as two international environmental regimes. The blind quest for normative hierarchy using legal techniques can lead to a sense of despair among lawyers studying regime interactions in similar policy domains. For example, in examining the case of

---

85 Koskenniemi 2005, 607; Khrebtukova 2008, 67–71.
86 Khrebtukova 2008, 63.
87 Michaels and Pauwelyn 2011.
88 Michaels and Pauwelyn 2011, 36.
89 Ghouri 2012.
90 Ghouri 2012, 268.

the climate and biodiversity treaties, Pontecorvo laments that 'there seems to be no way to establish a hierarchical order among the treaties involved in the conflict, neither according to their content nor according to their purpose'.[91] Conflict resolution rules such as *lex posterior* or *lex specialis* provide the lawyer with a seemingly objective standard, reflecting the will of states, to determine which norm prevails. However, states may not necessarily want one of the treaties' norms to prevail at the expense of another if there are potential synergies to be captured.

Such reasoning leads Wolfrum and Matz to argue for a 'forum for a common decision-making process' to pursue the harmonization of environmental treaties.[92] However, answering the forum question does not answer the question of what should guide the harmonization of two divergent treaties in such a way that the decision-making process is indeed 'common'. In other words, which objectives, principles or norms could guide decision making in such a way that it contributes to the objectives of both the interacting regimes?

There is a strong argument for using the concept of 'sustainable development' as an overarching objective for international environmental law and perhaps even international law more generally.[93] Christina Voigt, for example, argues that '[t]he classification of sustainable development as a general principle of law is legitimized by its widespread use in many national legal systems and in international law, and the jurisprudence of international courts and tribunals'.[94] One could argue, however, that the generality of this concept and the uncertainty about its precise contents or legal status reduce its usefulness. Criticizing the argument that sustainable development has become a binding norm of international law, Vaughan Lowe also sees potential for the concept to address the relationship between different norms. He argues that it constitutes 'a meta-principle, acting upon other legal rules and principles – a legal concept exercising a kind of interstitial normativity, pushing and pulling the boundaries of true primary norms when they threaten to overlap or conflict with each other'.[95] However, the risk is that such a concept simply becomes a new 'platform … through which experts representing the respective regimes may wage their struggle for influence'[96] rather than providing guidance for how to resolve conflicts.

---

91 Pontecorvo 1999, 740.
92 Wolfrum and Matz 2003, 158.
93 Voigt 2008a, 145–186.
94 Voigt 2008a, 186.
95 Lowe 1999, 31.
96 Koskenniemi 2012, 320.

Even if one does not accept that there are such all-encompassing objectives, principles or norms, or if one posits that their identification is impossible, the main point still holds true: legal techniques aimed at resolving normative conflict through normative hierarchy have overshadowed the idea that norms may also reinforce each other. The ILC has not completely ignored this critique. In its discussion of conflict clauses in international environmental law, the report acknowledges that in some cases it is necessary to use a clause that avoids 'a straightforward priority and instead seeks to coordinate the simultaneous application of the two treaties as far as possible'.[97] In particular, when treaties 'share a similar object and purpose or carry a parallel "ethos"', clauses aimed at mutual supportiveness are considered useful.[98] This is arguably the case for international environmental treaties, and is in line with the suggestion to formulate interaction clauses outlined above. The ILC then moves on to state that

> it cannot be assumed *a priori* that a similar readiness exists as between parties to treaties across regimes, treaties that seek to achieve physically incompatible solutions, or are inspired by very different (perhaps opposite) objectives in situations experienced as zero-sum games. In such cases, at the end of the day, one treaty must be preferred over the other.[99]

Whether 'such cases' exist in the area of global climate governance is questionable. The cases discussed in this book have highlighted only one potential normative conflict, the adoption of multilateral or unilateral climate-related trade measures that would violate WTO law. Under the ILC's view, if such a conflict were to materialize, it would be necessary to set aside one of the conflicting norms through 'regime-independent dispute-settlement'.[100] My first response would be that it is rather unlikely that cases involving conflicts between the climate, biodiversity and trade regimes would be brought before such regime-independent dispute settlement (e.g., the International Court of Justice or the Permanent Court of Arbitration). More importantly, if one accepts that the climate regime and related regimes are actually pursuing common goals, this whole quest for normative hierarchy becomes futile. The assumption of common goals may be more pertinent for international 'environmental' regimes, but it may also be possible to identify common goals in

---

97 ILC 2006, para. 272.
98 ILC 2006, para. 277.
99 ILC 2006, para. 277.
100 ILC 2006, para. 277.

the case of the overlapping international environmental and economic regimes. The potential synergies identified in Chapter 8 at least show that there are clear ways in which the two regimes can be mutually supportive.

4.3 Institutional Coordination: Opportunities

Bureaucracies

For constantly evolving regimes, bureaucracies can play an important role.[101] This role could also extend to interaction management. In most international environmental regimes, liaising with other secretariats is a task assigned to a treaty secretariat. This is also the case for the UNFCCC and CBD secretariats. Although cooperation is not explicitly included in the mandate of the WTO secretariat, it has liaised with other secretariats in practice.

As outlined in Chapter 5, the influence of secretariats in interaction management can be characterized as 'cognitive', 'normative' and 'executive'. The previous chapters show that in interaction management in global climate governance, cognitive influence can be discerned to some extent (corresponding to Oberthür's 'enabling' mode of interaction management[102]).

Institutional coordination between the climate secretariat and other bureaucracies has mainly concerned observership, mutual attendance at meetings, scientific cooperation, and information exchange, but these activities provide an important contribution to raising awareness about the existence of interactions, as well as their (potential) consequences.

The active role adopted by the CBD secretariat in raising awareness of the climate–biodiversity overlap, and the potential consequences for biodiversity protection of implementing climate measures, underlines the cognitive influence of bureaucracies in interaction management. While the secretariat received instructions for cooperation from the CBD COP, it also played an important informal role, as discussed in Chapter 7.[103] Whether the actions of the CBD secretariat extend to 'normative influence' remains unclear. While CBD COP decisions on climate change build on the preparatory work by the secretariat, it cannot be ascertained whether these decisions were directly influenced by this work.

---

101 Biermann and Siebenhüner 2009b.

102 Oberthür 2009, 377.

103 Jinnah 2011.

The formation of the Joint Liaison Group is another interesting development as it highlights the importance of information exchange and coordination between the administrative bodies of the different regimes.[104] Moreover, should parties decide so, it provides a potential forum for implementing measures that capture synergies between the three Rio Conventions. However, as Chapter 7 showed, this potential remains unfulfilled.

The WTO secretariat has also sought to exert cognitive influence by becoming active in enhancing the transparency of the WTO activities related to climate change,[105] for instance through the organization of events at COPs and the publication of reports on the linkages between trade and climate change.[106] And while institutional coordination between the climate secretariat and the APP was practically non-existent, Chapter 6 also showed that the secretariat has cooperated with organizations in the area of clean technology cooperation.

There are opportunities for enhancing the role of the secretariats in promoting synergies between different regimes. For instance, the tool of Memoranda of Understanding/Cooperation (widely used, for instance, by the biodiversity secretariat) has hardly been used by the climate secretariat. As noted in Chapter 5, secretariats arguably have 'implied powers' to conclude such memoranda with respect to overlapping issues that are important for the effective implementation of the treaty. Such written agreements could formalize existing informal practices, thereby keeping the relationship with other regimes permanently on the agenda of treaty bodies. However, it can be questioned whether formalizing cooperation per se would result in synergies at the operational level. Indeed, one of the advantages of the secretariats' activities is that they avoid the cumbersome political decision-making processes of the COPs and thereby provide a valuable informal and flexible way of integrating environmental regimes.

Still, there may be advantages to formalizing coordination in some instances. Chapter 6 highlighted various reasons why formalized coordination may be sensible in the case of the UNFCCC and minilateral initiatives. These reasons can be traced back to the closely related objectives of these institutions. Formalizing coordination could in such cases help avoid inefficiencies (in the form of duplication of work), and clarify that the institutions are meant to complement each other.

---

104 Chambers 2008, 69.

105 Jinnah 2010, 68.

106 UNEP and WTO 2009.

Decision-making bodies

The decision-making bodies of international regimes are also often instructed by treaties to cooperate with other bodies. This includes not only the climate treaty bodies and the CBD COP but also the General Council of the WTO. These mandates form the basis for the cooperation between the secretariats mentioned above, but also provide scope for the decision-making bodies themselves to engage in institutional coordination.

Not every decision-making body is equally active, however. In particular, the UNFCCC COP has been mostly silent about its relationships with other regimes. As noted in Chapter 7, only one COP decision was adopted on institutional cooperation, employing general and hortatory language.[107] In contrast, decision-making bodies of other international environmental regimes have sought to address the interactions with the climate regime. Chapter 7 provided an indication of the range of decisions adopted by the CBD COP related to climate change and biodiversity.[108] While these decisions have been important for raising awareness of the interactions, and have led to highlighting of biodiversity concerns in UNFCCC decisions,[109] they have not necessarily resulted in stronger biodiversity protection. More examples exist beyond the cases examined in this book. For instance, parties to the Montreal Protocol have been engaged in improving the relationship with the climate regime, most importantly by adopting a decision in 2007 that significantly accelerated the phasing out of the consumption and production of hydrochlorofluorocarbons, which are potent greenhouse gases that also served as a substitute for ozone-depleting substances.[110]

While institutional coordination on climate-related overlaps between regimes thus mainly takes place unilaterally, initiated by several proactive decision-making bodies, enhanced institutional coordination could take the 'more ambitious form of comprising joint planning of programmes or even the coordination of substantive decision-making or implementation activities'.[111] Examples of such enhanced coordination already exist in international environmental governance, for instance, in biodiversity

[107] UNFCCC Decision 13/CP.8, preamble.

[108] Morgera 2011.

[109] Yamin and Depledge 2004, 523–524.

[110] Oberthür *et al.* 2011. A similar decision to limit the use of another substitute with global warming potential, hydrofluorocarbons, has been proposed by parties to the Montreal Protocol, but is still opposed by others.

[111] Stokke 2001, 12.

protection, fisheries management and chemical substances.[112] Although extending the experiences from these issue areas to the climate regime may sound attractive in theory, there are limitations to what is possible and desirable, as will be discussed in the next section.

4.4 Institutional Coordination: Limitations

Unclear and restricted mandates

Although institutional coordination between the climate regime and other regimes is intensifying, its effects are uncertain. Institutional coordination can create mutual awareness and build capacity at various levels, but it is often also plagued by rhetoric about the 'mutual supportiveness' of different treaties, and devoid of practical suggestions. One of the underlying reasons is that institutional coordination is challenging because of unclear or restricted mandates.

Secretariats initiating cooperation with other bodies usually act upon a decision by the COP, thereby interpreting the mandate provided in such a decision. While it may seem 'commonsensical that a secretariat would not engage in activities against the will of its member states', it is not always easy for bureaucracies to ascertain what the 'will of the parties' is. In other words, secretariats do not always have a clear legal authority regarding the extent and contents of institutional coordination.[113] This may constrain or enable them. Parties will generally tend to interpret the secretariats' mandates restrictively, and secretariats will need to walk on eggshells when engaging in activities with other international actors. This is the case especially for the climate secretariat, which has been said to be 'living in a straitjacket' imposed by the parties.[114] The activities of the WTO secretariat are also closely scrutinized, as the reaction to the WTO-UNEP report on trade and climate change illustrated. However, other secretariats have taken a more proactive stance by adopting a wide interpretation of their mandate. The CBD secretariat has made use of the limited space provided to it by the COP to 'market' the linkages between climate change and biodiversity, partly due to a very active Executive Secretary.[115]

Cooperation between secretariats is even more difficult if their respective mandates differ in their scope. In the case of the climate–biodiversity overlap, this was exemplified by participants in the Joint Liaison Group,

---

112 Scott 2011, 202–208; see also Caddell 2012.

113 Chambers 2008, 70–71.

114 Busch 2009, 225.

115 Jinnah 2011.

which noted how the various Rio Conventions provided different signals to their bureaucracies, making it uncertain to what extent the forum could actually fulfil its mandate.[116]

Legitimacy and accountability

The mandate for cooperation and how it is interpreted will depend to a significant degree on parties' willingness to construct linkages with other regimes. This brings us to one of the core challenges to enhancing institutional coordination: countering the risk that states 'may be unwillingly drawn into regimes that they are not party to or affiliated with, and implicitly become subject to obligations under those regimes, by virtue of cooperative arrangement'.[117]

Any effort by actors in one regime to influence the normative development in another will likely be limited by the extent to which memberships are congruent. Chapter 7 highlighted this constraint for the climate–biodiversity overlap by emphasizing how the US (a party to the UNFCCC, but not the CBD) has responded to building bridges to the CBD. But even when membership is largely overlapping or, at the very least, the most powerful states are party to both the interacting regimes, there may be resistance to the idea of enhanced cooperation between bureaucracies.[118]

More generally, cooperation between the bodies of two different regimes gives rise to legitimacy and accountability concerns.[119] If one adopts a more traditional legal perspective emphasizing state consent (and state sovereignty) in international lawmaking, it is indeed difficult to see where the legitimacy of enhanced institutional coordination stems from, particularly in the case of incongruent memberships. A distinction should be made between situations in which institutional coordination takes place through bureaucracies and situations in which decision-making bodies are involved. States are not directly involved in the former, and thus cannot give their direct consent to institutional coordination whereas in the latter case they do have a say. The differences can be subtle. For example, if the UNFCCC secretariat drew upon the CBD secretariat to inform the definition of biodiversity safeguards, this could be regarded as the body overstepping its regime boundaries. By contrast, if the UNFCCC COP requested the CBD to become involved in defining

---

116 Chambers 2008, 69.
117 Scott 2011, 212.
118 E.g., Cossy and Marceau 2009, 376.
119 Scott 2011; M.A. Young 2011b, 271–284.

safeguards it would be more readily apparent that states had consented to institutional coordination.[120]

Concerns about legitimacy and accountability can be related back to the 'structural bias' of each regime.[121] If such a bias is indeed present, it is questionable whether institutional coordination can take place in a fashion that gives equal weight to the norms of each regime, particularly if institutional coordination involves a 'stronger' and a 'weaker' regime. In such cases, institutional coordination could result in the prioritization of one regime over another, meaning that coordination 'may become dominated by procedures, principles and concepts that are prevalent within one regime at the expense of [others]'.[122] Another matter is whether the norms of each regime *should* be given equal weight. Also in this case, Margaret Young argues that bodies seeking to cooperate with other regimes should 'scrutinise and review the "sources" of external regimes'[123] in a similar way that dispute settlement bodies should examine the origins of extraneous norms they may use. This scrutiny can focus on how procedural aspects, such as transparency and participation, have been respected in the other regime.[124] This means, for instance, that if the development of the other regime is characterized by a multilateral process that is conducted in an open and transparent fashion, norms developed under that regime would carry more weight than ones developed under a minilateral process closed to observers. This approach does not rely primarily on formal state consent but instead emphasizes consensus. Its appeal is that it can simultaneously make institutional coordination feasible and accountable, and thereby it avoids the risk of 'managerialism'.

4.5 Collective Interaction Management in Global Climate Governance

Having provided this overview of the opportunities and limitations of legal techniques and institutional coordination as a means of managing regime interactions in global climate governance, it is possible to see how these types of collective interaction management compare.

In terms of the consequences of interactions, legal techniques are (perhaps obviously) more relevant for interactions that involve or may

---

120 Cf. Matz-Lück 2012, 230.
121 Koskenniemi 2005, 600–615.
122 Scott 2011, 213.
123 M.A. Young 2011b, 277.
124 M.A. Young 2011b, 278–279.

involve normative conflicts. The usefulness of legal techniques based on the law of treaties to address regime interactions involving soft law or triggered by treaty body decisions is very limited. Ideally, conflict resolution techniques would resolve normative regime conflicts by reference to a higher standard for identifying which norm should prevail. However, I have indicated that it is hard to point to any normative conflict as a result of regime interactions in global climate governance in the first place. Moreover, even if such a normative conflict is identified, it is not necessarily desirable that one norm prevails over another. Legal techniques aimed at conflict *avoidance* circumvent some of these pitfalls, drawing attention in particular to the potential of harmonious treaty interpretation and carefully drafted interaction clauses.

In addition, legal techniques are more prominent when interaction management takes place through dispute settlement. Looking back at the previous chapters, such a situation is most likely in the case of a conflict between international environmental regimes and international economic regimes. Interactions that involve the climate regime and another regime in the same or a closely related policy domain are unlikely to be addressed before a dispute settlement mechanism. Not only is it unclear where a potential conflict involving the climate regime and, for instance, the biodiversity regime would be adjudicated in the absence of an international environmental court, the adversarial nature of litigation is also unlikely to attract much support from parties in both regimes and is generally ill-suited for international environmental disputes.[125] Of course, applying legal techniques such as harmonious treaty interpretation is not the sole prerogative of international adjudicators. However, it is clear that any balancing act undertaken, for instance, by the WTO Appellate Body will be more in the spotlight than the decisions of government officials in individual countries.

The prominence of legal techniques in dispute settlement does not necessarily mean that in such cases legal techniques will also be more successful or effective in managing interactions. This all depends on *how* the international adjudicator decides to apply the techniques. As discussed in Chapter 8, while the WTO dispute settlement bodies have applied legal techniques to manage interactions with international environmental law, they have done so arbitrarily.[126] Following the suggestions by the ILC and Margaret Young, there is potential for adjudicators linked to individual regimes, such as the WTO, to reach out to other areas of

---

125 E.g., Chayes and Chayes 1995, 205.
126 Kulovesi 2011b.

international law by scrutinizing the origins and, hence, the legitimacy and accountability of extraneous rules before taking them into account.

There are some indications of which factors could make institutional coordination successful. First, I argued in Chapter 6 that *formalized coordination* would ensure a better division of labour between the climate regime and minilateral clean technology agreements that could lead to efficiency gains. Second, in Chapter 7 I highlighted the potential benefits of *more structural coordination* in international environmental governance by discussing options for further integration of the UNFCCC and the CBD. Similarly, in Chapter 8 I argued that more structural coordination (in the form of a more permanent institutional dialogue) could help build trust and prevent the adoption of discriminatory trade measures. A structural form of coordination could help ensure that one of the likely conditions of successful interaction management, shared knowledge and understanding,[127] is fulfilled (see Chapter 5). Third, I indicated in Chapter 7 that enhanced *clarity of the legal mandate* for bureaucracies could assist secretariats in cooperating with other bureaucracies and international organizations, which could also help to enhance a shared understanding of the regime interactions. Fourth, the discussion above on the opportunities and limitations for institutional coordination suggests that coordination *involving the decision-making bodies* rather than bureaucracies is likely to be more effective, as it would most likely address the concern that states do not enter into obligations through the back door. Finally, although not directly flowing from the case studies, it could be speculated that *joint institutional coordination* (i.e., coordination involving treaty bodies from both interacting regimes) is likely to be more successful in managing the consequences of regime interactions. Even though the CBD's treaty bodies have been very active in highlighting the climate–biodiversity overlap, this activity has found its way only to a limited extent into the outputs of the climate regime. It is unclear, however, whether this would have changed if the UNFCCC's treaty bodies had engaged more with the CBD.

It is not certain that coordination efforts meeting the above-mentioned criteria would necessarily lead to better outcomes. Formalizing institutional coordination, as noted above, risks a return to a politically cumbersome decision-making process. And while some structure may be necessary to ensure continuity of institutional coordination, the creation of a permanent body as such does not necessarily lead to better shared understandings of regime interactions and ways of addressing them.

---

[127] Oberthür and Stokke 2011a, 324.

Moreover, given the 'formidable counterforces'[128] against more structural forms of institutional coordination, it is doubtful whether this option is at all politically feasible. The existence of a clear legal mandate is also not always enough for dealing with interactions if this mandate restricts rather than enables the treaty bodies. Finally, activities involving secretariats often result in more practical recommendations to address interactions than do the decisions produced by intergovernmental bodies.

Institutional coordination may also be more difficult in cases of intentionally triggered interaction resulting in a conflict.[129] As Young argues, it may be the case that 'the relevant players are likely to expend their time and energy devising ways to assert or enhance the dominance of their preferred approach to governance over the system(s) that others prefer'.[130] In the early stages of the APP, it would indeed have been hard to imagine that the US and Australia would be interested in formally coordinating their activities with the UN climate regime, seeing how the initiative was established with a view to showcasing a different model for international climate policy. This barrier to interaction management is related to what Oberthür and Stokke term 'political saliency':[131] if there is no political willingness to manage the outcome of regime interactions, interaction management is less likely to be successful.

4.6 Autonomous Interaction Management in Global Climate Governance

The case studies show that there are several regime interaction management options available that do not involve resorting to legal techniques or institutional coordination (i.e., that do not involve collective interaction management). It may even be argued that autonomous interaction management forms the primary means of interaction management in practice, as it seems to be less plagued by the limitations of legal techniques and institutional coordination outlined above. Chapter 6 first showed how Japan sought to manage its overlapping commitments to the APP and the Kyoto Protocol by adopting an ambivalent position towards the Partnership. Chapter 7 then highlighted the potential of enhancing the biodiversity benefits of forest carbon mitigation activities in the implementation phases by both state and non-state actors. Chapter 8, lastly, showed that in the absence of institutional coordination on BAMs, it would still be

[128] Oberthür and Stokke 2011a, 317.
[129] Zelli 2009, 64.
[130] O.R. Young 2008, 137.
[131] Oberthür and Stokke 2011a, 325.

possible to design the measures in a way that would respect the provisions of the climate treaties and the WTO agreements.

But do these autonomous efforts render legal techniques and institutional coordination obsolete? The short answer is 'no'. First, any autonomous interaction management by states party to the interacting treaties already implicitly uses the technique of interpretation. By adopting certain measures with a view to implementing the climate and biodiversity treaties simultaneously, parties (assuming they act in good faith and in line with the principle of *pacta sunt servanda*[132]) are interpreting the margin of discretion they have under both agreements. This re-emphasizes the importance of interpretation, and also underlines once again that interpretation is not only the task of judicial bodies but takes place on a regular basis by government officials and legal advisers.[133]

Second, autonomous management alone does not resolve underlying systemic tensions, and does not necessarily result in enhancing the effectiveness of both regimes simultaneously. As Oberthür argues, '[a]utonomous interplay management is least conducive for efforts aimed at systematically and structurally improving inter-institutional influence in [international environmental governance]'.[134] Ingrid Visseren-Hamakers and colleagues similarly conclude that the public–private partnerships they examined in the area of climate–biodiversity linkages 'have not prompted larger-scale structural improvements of existing negative and/or neutral influences between the climate change and biodiversity governance systems'.[135] In other words, autonomous management efforts do not address the relationship between different treaties in the long term.

Notwithstanding these critical remarks, autonomous interaction management can provide invaluable experiences that could help determine how two regimes could coexist in their implementation phase. The question is how such experiences of governmental and non-governmental actors could be integrated with intergovernmental efforts, and thus provide a foundation for mutually supportive regimes in cases where no structural solutions are readily available.

---

132 Pontecorvo 1999, 741–742.
133 Cf. Matz-Lück 2012, 213–218.
134 Oberthür 2009, 376.
135 Visseren-Hamakers *et al.* 2011, 102.

5. CONCLUSION

This chapter has sought to provide a cross-cutting overview of the variety of regime interactions in global climate governance, as well as insights into the usefulness of different types of interaction management.

Before moving on to the conclusions, a few methodological observations are in order. First, I have distinguished between actual and potential conflicts and synergies. While this distinction may be useful in theory, guiding attention to which conflicts or synergies may already exist and which may be more important in the future, it is not always clear in practice. For instance, I argued that the policy conflict between the climate and trade regime already exists because the objectives of the two regimes are conflicting. However, given their broad nature, the conflicting objectives may also form the basis for further (potential) conflicts in the future, for instance, specific measures adopted by parties to the climate regime that might violate the trade law disciplines. It is difficult to single out, through the process of causal mechanisms, the individual interactions and to separate actual from potential outcomes. Second, interaction management, again in theory, provides a way of dealing with the outcomes of regime interactions by reflecting on, and acting upon, the interactions. But, as Stokke and Oberthür note, interaction management may 'occur in anticipation of [the effects of interaction], thereby co-constituting the original interaction'.[136] In such cases, it is also not always clear where the interaction ends and where management begins.

Keeping these caveats in mind, this chapter has discussed various types of regime interactions in global climate governance. With respect to the object of interaction, the chapter has shown how the case studies presented in Chapters 6 and 7 support the argument to extend the analysis of regime interactions involving soft law arrangements. Regarding the causal mechanisms of interactions, it has made clear how the three main causal mechanisms identified by Oberthür and Gehring can all be identified in the case studies examined in this book. Furthermore, it has highlighted that the case of the APP provides some evidence of intentionally triggered regime interactions, while noting that in general intentionality is difficult to establish empirically.

Turning to the consequences of regime interactions in global climate governance, it is notable that none of the cases examined in this book constitute an outright normative conflict. Only the case of the climate–

[136] Stokke and Oberthür 2011, 7.

trade overlap shows that there is a potential normative conflict, but this depends on further developments in the climate regime and/or actions by individual (or groups of) countries. Policy conflicts have been more prevalent in the three cases. These policy conflicts were based on diverging objectives, the use of different principles and concepts, and opposing economic incentives. At the same time, various actual and potential synergies between the UN climate regime and other regimes could also be observed. These were based on shared principles and concepts, shared supporting measures, common economic incentives and the potential for streamlined monitoring and reporting. Cases of inter-institutional learning have not been identified. While this may still occur in the case of minilateral clean technology agreements, the outcomes of such learning are hard to predict.

Table 9.3 provides an overview of the findings concerning the types and consequences of regime interactions examined in this book.

Table 9.3 Types and consequences of regime interactions

	UN climate regime and minilateral clean technology agreements	UN climate regime and CBD	UN climate regime and WTO
Object of interaction	Hard law/soft law	Hard law/hard law (soft law)	Hard law/hard law
Causal mechanism of interaction	– Cognitive interaction – Interaction through commitment – Behavioural interaction	– Behavioural interaction	– Behavioural interaction – Interaction through commitment
Intentionality of interaction	(Partly) intentional	Unintentional	Unintentional
Consequences of interaction	– Actual and potential policy conflict – Actual and potential synergy	– Actual and potential policy conflict – Actual and potential synergy	– Potential normative conflict and actual policy conflict

The chapter then moved on to point out that many of the tensions involving the climate regime cannot adequately be captured by traditional legal definitions of conflict. In particular, interactions involving soft law or treaty body decisions are unlikely to be resolved by means of the law of treaties. This limits the usefulness of many techniques discussed in detail by the ILC. Furthermore, the chapter has questioned whether application of such techniques leading to a normative hierarchy is desirable at all, especially in the case of conflicts between two regimes in the same policy domain, such as two international environmental regimes. Moreover, it is unclear whether any overarching objectives, which could guide a decision on normative hierarchy, can even be said to exist.

Institutional coordination can complement the formal legal techniques for managing regime interactions in global climate governance. The chapter has discussed how administrative and decision-making bodies in climate-related regimes have started to address overlapping issues, with a view to addressing conflicts and maximizing synergies. However, it is clear that there are also limitations as to what can be achieved through such means. Secretariats' mandates are not always clearly defined and to avoid a rebuke by parties, secretariats will tend to stay away from intruding too much into the decision-making process through external cooperation with other institutions. This question is linked with more general concerns about the legitimacy and accountability of institutional coordination. These concerns are to some extent based on traditional notions of state consent, but they point to the real risk that actors in one regime are sidelined through the use of norms borrowed from another.

Finally, the chapter has argued that autonomous interaction management may be useful in the absence of higher levels of interaction management. However, autonomous interaction management by states is actually not that different from harmonious treaty interpretation. Furthermore, autonomous interaction management as such does not address the overall relationship between two regimes.

10. Conclusions and ways forward

> [I]nterplay management often helps to ameliorate problems that many authors have associated with institutional interaction: fragmented and contested knowledge on overlapping issues …, duplication of effort …, and normative ambiguity due to conflicting commitments under separate institutions …[1]

> Managerialism is not a solution. It is a problem.[2]

1. INTRODUCTION

As a 'super wicked' problem with a causal chain marked by complexity and uncertainty, it is impossible to govern all facets of climate change through a single international regime. To be effective, the international climate regime will need to take into account the potential of a variety of international regimes to either mitigate or exacerbate the problem, while at the same time also considering its own impacts on other regimes. Global climate governance is thus inevitably fragmented. The consequences of such fragmentation, however, are not necessarily duplication of work, regulatory chaos and uncertainty, or the prioritization of one policy field over another, as has been feared by international lawyers. Crucially, I have argued in this book that the implications of the fragmentation of global climate governance ultimately depend on how regime interactions are managed.

In this chapter, I offer concluding observations that seek to explore the middle ground between the views of interaction management optimists, such as Sebastian Oberthür and Olav Schram Stokke, and sceptics like Martti Koskenniemi, who hold opposing views on whether interaction management can work in the first place. In addition, I outline the main policy implications for global climate governance that emerge from this study and reflect on areas for further inquiry.

---

[1] Oberthür and Stokke 2011a, 315.
[2] Koskenniemi 2012, 324.

2. CONCLUSIONS

2.1 Consequences of Regime Interactions

The debate on the fragmentation of international law and governance has highlighted various promises and pitfalls of fragmentation at the international level, as I outlined in Chapter 3. I have sought to go beyond this largely academic debate by adding new insights into the specific consequences of regime interactions in the area of global climate governance.

To this end, Chapter 4 first introduced a typology of consequences of regime interactions. This typology added two important nuances to the familiar and relatively straightforward distinction between 'conflicts' and 'synergies'. First, I argued that for conflicts it is useful to distinguish between a narrow legalistic definition of 'normative conflicts' and a broader definition of 'policy conflicts'. This distinction captures the fact that from an international lawyer's point of view not all tensions between regimes should be regarded as 'conflicts', while at the same time conceding that those broader tensions also warrant attention. Second, I suggested that some consequences of regime interactions may already have manifested themselves (i.e., they are 'actual') whereas in other cases consequences may not yet have materialized (i.e., they are 'potential'). Beyond introducing this typology, the chapter also provided guidance on how conflicts and synergies could be identified. It pointed to the methodological challenges this identification entails, as notions of 'conflict' or 'synergy' are strongly related to often-ambiguous standards of regime effectiveness. Acknowledging these limitations, the chapter proposed several indicators at the regime output level that allowed for the identification of a potential or actual conflict or synergy in the subsequent case studies. While falling short of a theory-based explanation of *why* regime interactions entail certain consequences, these indicators help build the argument that a particular consequence can be considered conflictive or synergistic.

The subsequent chapters applied this part of the analytical framework to three different types of regime interactions: (1) within the (narrowly defined) field of international climate law and governance, focusing on interactions between the climate regime and minilateral clean technology agreements; (2) within the field of international environmental law and governance, focusing on interactions between the climate regime and the CBD; and (3) between two distinct fields of international law and governance, focusing on the interactions between the climate regime and the WTO. The analysis of three different types of regime interactions not only offers insights into the diversity of interactions and their consequences in a particular issue area, but also allows for making more general inferences about how the various types of interactions relate to the types of consequences.

First, given the overlapping objectives it is reasonable to expect that interactions between two regimes in the same policy domain will primarily have positive outcomes. However, Chapter 6 showed that this need not be the case. Using the Asia–Pacific Partnership as an example, several potential policy conflicts between the climate regime and minilateral clean technology agreements were identified. While the different regimes may have largely similar goals (i.e., climate change mitigation), the pathways for achieving these goals diverged significantly in practice. I related these differences to the actors participating in the interacting regimes, with minilateral regimes seeking to limit participation to a number of key countries, as opposed to the UN climate regime's multilateral process. The differences also concerned the nature of commitments adopted under the regimes, with the APP falling more within the realm of soft law compared to the hard law of the UNFCCC and the Kyoto Protocol. These (and other) differences between minilateral clean technology agreements and the UN climate regime were at least in part traced back to the policy positions of key actors, notably the US and Australia, who launched the minilateral regime. As Chapter 9 underscored, the intentionality behind the creation of new minilateral clean technology agreements may well have led to the potential clashes with the UN climate regime. At the same time, Chapter 6 argued that the potential for conflict between minilateral clean technology agreements and the multilateral regime should not be exaggerated. None of the new minilateral venues have replaced the UNFCCC as the central forum for international climate policy (and are unlikely to do so in the near future), and the work carried out under them could contribute (and arguably has already contributed) to the effective implementation of the climate treaties. Finally, while concerns have been raised about the cognitive influence of the APP and other minilateral initiatives on the UNFCCC process, the jury is still out on how this would affect the performance of the latter.

As regards the second type of interaction, international environmental law is often seen as a special field of international law, characterized by common principles (e.g., the no-harm principle or precautionary principle), procedures (e.g., compliance mechanisms) and instruments (e.g., economic instruments). Consequently, it seems safe to assume that regime interactions within international environmental law would mainly lead to beneficial outcomes. However, the case study of interactions between the climate regime and the CBD showed that for this type of regime interaction as well, the consequences may vary. While the objectives and underlying values of the climate and biodiversity regimes may be pointing in the same direction, with both at least indirectly aiming to protect and conserve ecosystems, certain climate policy measures may still have adverse impacts on biodiversity. Chapter 7 focused on the role of forests as sinks and sources of carbon, showing how measures promoting the use of forests as carbon sinks under the Kyoto Protocol led to an actual policy conflict between the two regimes. The consequences of the forest-related regime interactions depend to a great extent on the economic incentives provided by the measures designed under the climate regime. Under the Kyoto Protocol, the safeguards for protecting biodiversity were not strong enough to counter the strong economic incentives provided by the carbon market, resulting in a policy conflict. Whether this situation will change for the UNFCCC's mechanism on reducing emissions from deforestation and forest degradation depends on the strength of the economic incentives as well as the operationalization of the safeguards, two aspects of REDD that remain under negotiation. Like the first case, Chapter 7 showed that overlapping objectives are insufficient to prevent (potentially) conflictive outcomes and to promote synergies. Diverging conceptual approaches and the use of different policy instruments may lead to policy conflicts between two regimes within the same field of international environmental law and governance.

The third type of regime interaction involves two more clearly distinct fields, international economic law and international environmental law. This type of regime interaction can be expected to result in more pronounced tensions, as objectives, principles and norms are likely to diverge. Indeed, the only potential normative conflict identified in the case studies involved the interactions between the climate and trade regimes. However, it should immediately be added that no such normative conflict has emerged as yet. This may of course change in the future, either with the adoption of a new international climate change agreement containing trade measures or, rather more likely, because of the adoption of unilateral climate-related trade measures. In addition to this potential normative conflict, Chapter 8 pointed to various policy conflicts between the trade and climate regime, seemingly reinforcing the long-standing perception that the fields of international environmental law and international trade law are at odds with each other. However, the analysis also identified several potential synergies between the trade and climate regimes (e.g., reducing fossil fuel subsidies and liberalizing trade in climate-friendly goods and services), showing that there evidently is room for the two regimes to support each other.

The case studies highlighted a wide variety of actual and potential consequences of regime interactions in global climate governance. One of the most striking findings is that despite the concerns raised in the debate on the fragmentation of international law, no normative conflict could be identified. This does not mean that there are no tensions between the climate regime and other regimes; the small sample of regime interactions covered in this book already provides sufficient evidence for broader policy conflicts. However, the absence of outright conflicts between norms urges us to look beyond methods and approaches aimed primarily at avoiding or resolving normative conflict. I return to this point below.

A second key finding is that there are various instances of synergy resulting from regime interactions in global climate governance. While this finding in itself is not novel and is in line with earlier general studies on interactions in international environmental governance,[3] it bears repeating in light of the continued preoccupation of the international law community with regime conflicts. I do not mean to wish away conflicts or suggest that this preoccupation is wholly unjustified. However, international lawyers still pay insufficient attention to the conditions under which certain regime interactions may result in synergy. These conditions are not only linked to the nature and structure of a particular problem, but they may also be related to design elements of a particular regime, implying that international lawyers have a role to play. This finding is clearly apposite for regime interactions in more narrowly defined fields (e.g., global climate governance; international environmental law), where different regimes are expected to contribute to the same goals. However, how and why regime interactions lead to synergies could also be explored in more depth in global governance more broadly.

Third, whether a regime interaction leads to conflict or synergy often depends on factors that are seemingly under the control of actors participating in the interacting regimes. For instance, the cognitive interaction identified in Chapter 6 on how the climate regime could learn from minilateral clean technology agreements requires some form of agency: these lessons will have to be learned by specific actors, be they individual parties, country coalitions or administrative bodies. To use a different example, Chapter 7 argued that economic incentives played an important role in steering the outcome of interactions towards conflict or synergy. What those economic incentives entail is not a given; it is parties to the climate regime that have the power and authority to decide what

[3] Gehring and Oberthür 2006a, 318.

the incentives will be, for example by choosing to link REDD to carbon markets or not. The point is that actors participating in the interacting regimes can influence the outcome of regime interactions through interaction management.

2.2 Management of Regime Interactions

In Chapter 4, I discussed various types of interaction management, with an emphasis on collective forms of interaction management (i.e., going beyond implementation by individual state and non-state actors). Drawing attention to the specific role of international law, I introduced a broad distinction between legal and political approaches in interaction management. The first category consists of various legal techniques well known to international lawyers, such as treaty drafting, treaty interpretation, conflict clauses and priority rules for the resolution of conflicts of norms such as *lex superior*, *lex specialis* and *lex posterior*. These techniques can not only be applied by judicial bodies, but may also be employed by negotiators and government officials at the domestic level. With the second category I refer to a range of activities carried out by (groups of) actors participating in the interacting regimes that do not involve legal means (e.g., dispute settlement). These actors include the regimes' decision-making bodies (e.g., the UNFCCC COP), administrative bodies (e.g., the CBD secretariat) or dispute settlement bodies (e.g., the WTO Appellate Body). I suggested that these actors can seek to influence the outcomes of regime interactions individually or jointly; on an ad hoc or a more permanent basis; and with or without a clear legal mandate to do so.

The mere fact that legal and political approaches are employed in practice is not sufficient to argue that they are *effectively* managing regime interactions. As I showed in Chapter 4, this determination is highly subjective and largely dependent on the existence of an overarching standard or yardstick against which to measure the effectiveness of interaction management. I therefore argued for a pragmatic approach in assessing the effectiveness of interaction management, focusing on (1) an enhanced shared understanding of regime interactions; (2) possible efficiency gains reaped; and (3) the avoidance or resolution of normative conflicts.

Applying this analytical framework, Chapters 6–8 examined how these approaches have been, and could be, applied to regime interactions in global climate governance.

The first case study illustrated that the potential for interaction management through legal techniques is severely limited in the case of soft–hard law interactions. This finding is not necessarily related to the fact that the two regimes are situated in the same policy field (i.e., global climate governance). However, at the international level, the more narrowly one defines a specific field, the likelier it is that there is only one 'core' hard law regime.[4] This means that soft–hard law interactions are thus likelier to be more prominent in narrowly defined issue areas, and that accordingly the scope for using legal techniques may become more limited in such cases.

The limited usefulness of legal techniques in addressing soft–hard law interactions redirects attention to the role of institutional coordination. Chapter 6 showed that a minimum level of coordination existed between the climate regime and minilateral clean technology agreements. This existing coordination was largely on an ad hoc basis, and for various minilateral initiatives (including the APP) there was no meaningful attempt to coordinate at all. This implies that it was difficult to foster a shared understanding of regime interactions between the climate regime and minilateral clean technology agreements or to benefit from efficiency gains over time by avoiding contradictions and duplication of work. While the UNFCCC and minilateral clean technology agreements each have an important role to play in promoting clean technology development and transfer, they have not yet reached a clear division of labour that makes use of their respective strengths. Ensuring that a variety of minilateral and multilateral regimes work towards the same (or at least similar) goals as effectively and efficiently as possible will therefore likely require enhanced institutional coordination. This argument may be extended to other specific issues in global climate governance where the UNFCCC's work is flanked by a wide variety of public, private and public–private institutions, including REDD, emissions trading, climate finance, short-lived climate pollutants and adaptation.[5]

The analysis of interaction management in the second case study revealed that in the area of international environmental law and governance, where it can be said that 'most participants will at least share basic

---

[4] For instance, if 'biodiversity' is the broadly defined issue area under scrutiny, the analysis would include a wide range of biodiversity-related treaties, including the CBD, the Ramsar Convention, the Convention on International Trade in Endangered Species, the Convention on Migratory Species, etc. However, if one delineated the issue area more narrowly to, for example, access and benefit sharing of genetic resources, the 2009 Nagoya Protocol on Access and Benefit Sharing would more clearly be the core (hard law) treaty.

[5] Cf. van Asselt and Zelli 2014, forthcoming.

assumptions about the values at issue',[6] the relevance of legal techniques is also limited. More specifically, one of the key characteristics of international environmental law, international administrative lawmaking through decisions by treaty bodies,[7] constrains the potential for legal techniques in dealing with conflicts. Both norm-based and policy-based conflicts are unlikely to result directly from the black letter texts of treaties, but could instead be triggered by the subsequent operationalization of treaty provisions through decisions, which are not clearly covered by the law of treaties. Nevertheless, the ambiguity and room for discretion often found in treaties also presents an opportunity for interaction management, as it opens up space for harmonizing two regimes through treaty interpretation. Moreover, the framework-protocol structure found in the UN climate regime (as well as in other international environmental regimes) offers opportunities for influencing the outcomes of regime interactions through ongoing negotiations after a treaty is adopted.

Related to the above-mentioned limitations of legal techniques, interaction management through institutional coordination seems particularly promising in the area of international environmental governance.[8] As I showed in Chapter 7, the activities of the CBD's treaty bodies in particular have enhanced shared understanding about the (potential) consequences of interactions between the climate and biodiversity regimes. This enhanced understanding has not removed all tensions between the regimes, but there are various ways in which coordination can be intensified. Many of these options have been discussed before in the context of the debate on the reform of international environmental governance[9] and, in the context of the Rio+20 summit, on the institutional framework for sustainable development.[10] Most suggestions raised in these discussions point to more structural forms of institutional coordination, at times combined with a stronger legal mandate (the proposal for establishing a World Environmental Organization being the most far-reaching example). However, as I argued in Chapter 9, these more structural forms of institutional coordination may not be feasible when membership of the interacting regimes differs. More importantly, the options that envisage a strengthened role for bureaucracies, including not only treaty secretariats but also bureaucracies established under new

---

6 Klabbers 2008, 94.
7 Brunnée 2002; Wiersema 2009.
8 Cf. Wolfrum and Matz 2003; Chambers 2008; Scott 2011; Caddell 2012.
9 E.g., Biermann and Bauer 2005; Fauchald 2010.
10 E.g., Bernstein and Brunnée 2011.

overarching institutional structures, will likely face an uphill struggle, as states will be wary of ceding (too much) decision-making authority. Of course, not all bureaucracies will be equally constrained. Bureaucracies are less likely to be able to exert autonomous influence (including through interaction management) if problems are characterized by high costs of international regulation, or if the issue is 'salient' (i.e., if the problem can be classified as 'environmental high politics').[11] Similarly, 'problem malignancy' and 'political saliency' are important factors in determining the success of interaction management.[12] The 'malign' nature of the climate problem provides a strong explanation for the constraints faced by the UNFCCC secretariat in interacting with other international environmental bureaucracies. High political saliency could also form a part of this explanation, but it may equally be the reason why institutional coordination takes off in some cases (see below).

Interaction management in the third case study focused particularly on the potential of legal techniques. This is not surprising, as the international trade regime is characterized by a strong and active dispute settlement system. The chapter discussed the legal techniques through which international climate change law could play a role in settling a possible dispute on climate-related trade measures. Following the attention it has received in the debate on the fragmentation of international law, the chapter focused on the potential of the 'principle of systemic integration' to harmonize international climate and trade law. It showed in particular how climate law and policy could be brought into a possible WTO dispute relating to climate measures through treaty interpretation. However, citing WTO case law, it questioned the extent to which international adjudicators connected to a specific regime are willing to apply this technique in practice and open up 'their' regime to extraneous rules.

In Chapter 8, I also argued that institutional coordination between the climate and trade regimes could reduce sensitivities and help build trust around the issue of climate-related trade measures, but I made clear that institutional coordination in practice has been limited. Also in this case, the problem malignancy of climate-related trade measures constrains the WTO secretariat's room for manoeuvre. Nevertheless, the high saliency of the issue in recent years has made it possible to at least initiate

---

11 Biermann and Siebenhüner 2009a, 334–335.

12 Oberthür and Stokke 2011a, 324–325.

institutional coordination,[13] leading to a basic level of shared understanding of the regime interactions. Although Chapter 8 identified several options for joint coordination by the UNFCCC's and WTO's decision-making and administrative bodies, the feasibility of these options is hampered by the differing ideologies underlying the regimes, as well as incongruent memberships. Still, unilateral institutional coordination by discussing trade measures under each of the regimes is possible. I exemplified this by discussing why and how institutional coordination could be initiated for the issue of border adjustment measures.

While the case studies in this book focus on one particular issue area in global governance, it is nevertheless possible to highlight broader findings relevant for the debates on institutional fragmentation and regime interactions.

First, in a world where the lines between law and non-law are blurry, and where international law based on formal sources is increasingly accompanied by informal international lawmaking,[14] the relevance of formal legal techniques and the law of treaties in addressing regime interactions is increasingly being challenged. While different regimes may be at odds with each other, the tensions between them cannot always be traced back to the texts of treaties upon which the regimes are based.

Second, legal techniques nevertheless hold potential for managing regime interactions if they are reconsidered in original ways. This book offers several suggestions. Treaty interpretation could be employed in a way that does not require parallel membership but instead directs adjudicators to critically examine the sources of extraneous norms. Moreover, while dispute settlement bodies will remain an important locus for applying the technique of treaty interpretation, it should not be forgotten that treaty interpretation also takes place at the domestic level, meaning that attention should also be paid to the role of national-level policymakers. Another well-known legal technique that could be re-invented to tackle regime interactions is the inclusion of provisions that clarify the relationship between treaties. Such 'interaction clauses' could go beyond 'conflict clauses' by: acknowledging not only potential conflicts between different treaties, but also potential synergies; clearer drafting, turning the clauses into something more than paper tigers; and mandating inter-institutional coordination, adding a dynamic and flexible element that has so far been lacking.

---

[13] Cf. Oberthür and Stokke 2011a, 325.

[14] Cf. Wessel 2011; Pauwelyn 2012.

Third, the general limitations of legal techniques require us to think more deeply about institutional coordination as a means of interaction management. I have offered initial insights into the advantages and drawbacks of different types of institutional coordination (formal/informal; structural/ad hoc; strong/weak legal basis; involving decision-making/administrative bodies). The case studies further provided suggestions regarding which types of institutional coordination may lead to efficiency gains or enhance shared understanding of regime interactions. Taken together, however, they are still inconclusive about which type of institutional coordination works best in which situation.

Fourth, although institutional coordination may be feasible in certain cases, it also raises key questions of legitimacy and accountability, as interaction management may sideline the interests and concerns of actors in the interacting regimes.[15] The risk that institutional coordination by treaty bodies does not reflect the consent of the states participating in the interacting regimes is valid in particular for bureaucracies, which generally are not granted decision-making authority. Moreover, when two interacting regimes can be considered unequal, for instance because powerful states participate in one regime but not the other, institutional coordination may become a vessel for those states to wield their influence beyond a particular regime. One way to address these accountability and legitimacy concerns is for treaty bodies wishing to engage in institutional coordination to carefully review how the norms in another regime have been created, and whether this has been done in a process that enjoyed wide participation and can be characterized by transparency and openness.

Fifth, autonomous interaction management can complement collective forms of interaction management in important ways, but that is in itself insufficient as a means to effectively address regime interactions. The main reason for this is that autonomous interaction management does not address underlying systemic tensions in international law and governance. International regimes can generally be considered to be more durable than policy positions adopted by individual governments and even though actions by non-state actors can pave the way for effective interaction management at the international level, they still require scaling up.

Lastly, there are no clearly identifiable and objective standards against which interaction management can be evaluated. This is not for a lack of suggestions. For instance, some conflict resolution techniques are linked

---

[15] M.A. Young 2011b, 271–284; Scott 2011.

to the rationale that they can resolve a conflict by sticking as closely as possible to what states actually intended. Hence, under *lex specialis*, the idea is that a more specific rule better reflects states' intentions than a general one; and the suggestion of *lex posterior* is that the later rule provides the most up-to-date reflection of what states want. Another reaction from the international law community has been to search for hierarchically superior norms (*lex superior*). And while the concepts *jus cogens* or obligations *erga omnes* provide lawyers with tools to identify such norms, they inevitably lead to the follow-up question: what are the qualifying criteria for such norms? Seemingly all-encompassing concepts such as sustainable development have also been suggested. However, the largely abstract nature of this concept and its often contested meaning imply that it, like other concepts that try to identify 'what constitutes the good life',[16] cannot provide sufficient guidance for dealing with very specific regime interactions. Such concepts may thus end up as a 'regime hybrid' comprising the vocabulary of different regimes seeking to influence each other.[17] More generally, the laudable goals often pursued in managing regime interactions ('harmonization', 'mutual supportiveness', achieving 'coherence' or 'balancing' interests) will always lead to an outcome that favours some interpretation over others.[18] In short, 'attempts to construct a system of "constitutional" rules governing the relative priority to be accorded to different bodies of international law' have not 'been clearly successful, either in the sense of producing clear conflicts rules, or in the sense of producing an institutional and social space for the productive discussion of the values questions which underlie inter-regime conflicts'.[19]

But does this mean, then, that all interaction management is in vain and that Koskenniemi is right in positing that '[i]n a world of hegemonic regimes, there is no "innocent" or impartial, neutral terrain from the perspective of which regime interaction could be managed'?[20] As should be clear by now, I do not think so. More specifically, Koskenniemi is correct in asserting that interaction management cannot resolve the clashes between values and interests that underlie some regime interactions; it provides a way forward for what Jan Klabbers terms 'co-ordination problems'.[21] There are two key points to make here. First, not

---

16 Klabbers 2008, 100.
17 Koskenniemi 2012, 319–320.
18 Lang 2012, 113.
19 Lang 2011, 147.
20 Koskenniemi 2012, 320.
21 Klabbers 2008, 18.

all regime interactions reflect ideological divides. To be sure, this book has discussed various instances where they do, but in some cases interaction management is clearly feasible. Examples can be found in the quotation of Oberthür and Stokke with which I opened this chapter. Through interaction management, it is possible to enhance a shared understanding of how specific policy problems, and the institutional frameworks governing them, are related. Moreover, in some cases, linking implementing activities can help reduce inefficiencies and allow for inter-regime learning. Second, even if values do clash, there is still a sound argument for the type of accountable interaction management mentioned above. Hence, in regime interactions characterized by high stakes (e.g., the climate and trade regimes), there is still a strong case for the interacting regimes to take extraneous norms into account as long as the sources of those norms are carefully reviewed.

Even scholars emphasizing the benefits of multiple legal orders have argued for some form of interaction management. As legal pluralist Gunther Teubner suggests: 'If we lift our gaze from simple division of labour to the more complex functional differentiation of society, we can then see that the high autonomy of global function systems demands a new type of sustainability and a new sensitivity for their environments.'[22] Similarly, Nico Krisch, in a strong defence of pluralism in 'postnational law', argues that 'interface norms' are needed to coordinate different legal systems.[23] Perhaps, then, in the absence of 'constitutional rules' for dealing with regime interactions, it is possible to think of interaction management through coordination as a rather modest effort on the road to international constitutionalization.[24]

3. IMPLICATIONS FOR POLICYMAKING

Although the focus of this book has in part been on academic discussions regarding the fragmentation of international law and regime interactions, several aspects are relevant from a policymaker's perspective. Below, I discuss the main policy implications for each of the case studies. In addition, I indicate how the role of the climate regime could evolve in an increasingly fragmented and complex institutional environment.

---

22 Teubner 2012, 173.
23 Krisch 2010, 285ff.
24 Dunoff and Trachtman 2009a, 30–31.

3.1 The UN Climate Regime and Minilateral Clean Technology Agreements

States have increasingly created minilateral regimes (or started using existing ones) in the fight against climate change. A key idea behind the minilateral approach is that it is easier to get to agreement among a smaller number of like-minded countries (i.e., a 'coalition of the willing'[25]) than through a multilateral negotiation process. Moreover, by departing from the intergovernmental structure of multilateral regimes, minilateral initiatives could clear the way for innovative governance approaches involving the private sector.

Although this logic sounds appealing, it is not entirely convincing. If the small group of countries overlaps largely with the countries causing a stalemate in the global negotiations (notably the world's largest emitters), then it is unlikely that minilateral approaches will fare better than multilateral ones.[26] Moreover, the exclusion of a large group of countries inevitably raises concerns about the legitimacy of the decision-making process.[27] The multilateral process guarantees that less powerful, vulnerable communities that do not have a voice in minilateral initiatives are represented and that the final outcome reflects their concerns and interests. Finally, the creation of minilateral alternatives to the UNFCCC could simply be seen as an instance of forum shopping, with powerful states choosing to discuss international climate policy in the venues that best suit their interests.

Yet ultimately, this 'either/or' debate of whether multilateral or minilateral regimes are the best way forward ignores the reality of global climate governance: they already coexist.[28] It is for this reason that I sought to draw attention to the ways in which minilateral clean technology agreements could coexist and cooperate with the multilateral climate regime.

The starting point here is the strengths and weaknesses of both multilateral and minilateral regimes. On the basis of the identified strengths and weaknesses it is possible to suggest a division of labour between minilateral and multilateral regimes. As I argued with respect to the issue of technology development and transfer, the UNFCCC can continue to play an important role in identifying developing countries' climate technology needs, and has the potential to link technology

---

25 Hale 2009.

26 Depledge and Yamin 2009, 451; Eckersley 2012, 33.

27 Eckersley 2012; Karlsson-Vinkhuyzen and McGee 2013.

28 Cf. van Asselt and Zelli 2014, forthcoming.

initiatives to various funding mechanisms. Minilateral clean technology agreements, conversely, have an important role to play in the design and implementation of specific actions related to technology development and transfer. While this division of labour could emerge implicitly, I argued that institutional coordination, through more frequent and comprehensive reporting by the administrative bodies of minilateral agreements to the multilateral climate regime, could enhance transparency and predictability by showing how the actions of other institutions contribute to the UNFCCC's objective. This coordination could be integrated in the work programmes of new technology bodies such as the Technology Executive Committee and the Climate Technology Centre and Network, which could in turn report to the COP.

3.2 The UN Climate Regime and the Convention on Biological Diversity

Forests play a crucial role in tackling the twin problems of climate change and biodiversity loss. However, so far a multitude of international efforts has failed to provide adequate protection of the world's forests. For some, the emerging issue of REDD seems to provide a promising avenue, as it could provide developing countries (and key stakeholders within those countries) with a financial incentive to protect their tropical forests. However, while REDD may lead to significant climate mitigation benefits, its biodiversity impacts vary depending on design elements that still require international consensus.

I have not sought to provide an ideal solution to managing the interactions between the climate and biodiversity regimes, for instance by giving design recommendations for a REDD mechanism. While biodiversity is clearly an important issue for successful REDD design, it is but one of the many issues that international policymakers need to tackle (the rights of indigenous and local communities being another crucial issue). Moreover, although I focused on the nexus of two specialized regimes, it should be made clear that there is a plethora of public, private and public–private initiatives at various levels of governance, which all work in interaction with each other. A full assessment of the effectiveness of global forest governance will need to cover this entire institutional complex (see below).

Bearing in mind these limitations, Chapter 7 showed that there was potential to reap synergies between the climate and biodiversity regimes through collective interaction management in the form of enhanced institutional coordination and autonomous interaction management by states and non-state actors.

With respect to collective interaction management, I suggested that it may be feasible to agree on a short reference to the contribution of REDD to biodiversity (or other) benefits in a future legally binding agreement, a provision which could open up negotiating space for better integration of biodiversity concerns in the future. This would mean that the broad obligation to protect biodiversity would be anchored in a treaty, but that the manner in which this was to be achieved would be postponed to a later stage. Various options to integrate biodiversity considerations would then still be available, but would need to be strengthened by using more precise or more mandatory language in such decisions. This would provide an opportunity to reinforce the inclusion of biodiversity considerations in the UN climate regime over time.

Given that the climate and biodiversity context of forest projects will vary country by country, detailed and stringent common biodiversity standards for forest carbon activities adopted either under the UNFCCC or under the CBD will likely be unacceptable (as well as undesirable). However, there is potential for further collaboration on monitoring and reporting of biodiversity impacts. Over the years, the CBD has built up significant institutional capacity in the area of monitoring and reporting on various aspects of biological diversity, and it is sensible to build on its experience.

As regards autonomous interaction management, I emphasized the potential for both states and non-state actors to ensure that climate policies in the forestry sector result in biodiversity benefits. Governments in both developed and developing countries can play a role here. Developed countries can seek to safeguard biodiversity through financial channels, either through direct funding to projects in developing countries or through market access requirements. In both cases, they can link the pledging or delivery of financial resources to the achievement of biodiversity benefits. In addition, developing countries can adopt policies and measures, such as domestic biodiversity standards, impact assessment requirements, and generally improving policy and legal coherence at the national and sub-national levels. Efforts by non-state actors could support these actions, for instance by providing services in terms of monitoring, reporting and verifying the biodiversity impacts of climate policies.

3.3 The UN Climate Regime and the World Trade Organization

There is a wide variety of interactions at the intersection of the climate and trade regimes. While I do not provide immediate suggestions on how to deal with each of these interactions, Chapter 8 provides new insights into how to tackle one of the most politically sensitive issues: the use of unilateral trade measures.

First, I argued that there is no single forum to exclusively discuss the use of climate-related trade measures. Put simply, if discussed in the WTO context, there will be concerns related to the (lack of) environmental credentials of the organization. If discussed under the UNFCCC, there will be concerns about disguised protectionism. If discussed outside of both regimes, there will be concerns that any agreed solution will still need to be fed back into the trade and climate negotiations. Therefore, I suggested pursuing discussions simultaneously under the umbrellas of the UNFCCC and the WTO. Both regimes already provide a forum for their discussion. Under the UNFCCC's agenda item of 'response measures', various proposals on how to deal with unilateral trade measures have already been discussed. And the WTO's Committee on Trade and Environment provides a non-negotiating setting for similarly discussing the advantages and drawbacks of climate-related trade measures.

Second, while the adoption of climate-motivated trade measures may seem laudable from the viewpoint of climate change mitigation, the risk of disguised protectionism is not imaginary.[29] Therefore, rather than waiting for a measure to be challenged before the WTO's dispute settlement bodies, it is important for states wishing to adopt a measure to make the climate change rationale as clear as possible in a non-adversarial setting, and to convincingly show that the measure is not simply designed to protect domestic producers or service providers. By drawing on the substantial body of scientific evidence on carbon leakage and competitiveness, the discussions on BAMs could thus move away from the rhetoric of protectionism versus climate protection. In addition, developed countries seeking to use trade measures would need to show how they seek to minimize the impacts on developing countries (especially least-developed countries) or how these countries may possibly benefit from trade measures. In other words, developed countries need to show exactly how they intend to respect the principle of common but differentiated responsibilities and respective capabilities. Again, this would need to be done on an informed basis, which includes examining the impacts of different design options on developing countries.[30] As the related case of the EU's inclusion of aviation emissions in the EU emissions trading scheme shows, the lack of fairness in unilaterally

---

29 E.g., Epps and Green 2010, 210.
30 E.g., Cosbey *et al.* 2012.

adopted climate measures has been a key source of contention for developing countries, stressing the need to address this concern.

Finally, in case unilateral trade measures are inevitable, it is useful to design them in such a way that they could be changed or made void if the climate objectives can be achieved through other means. The aviation case is again a good example: while the EU argued that it needed to include emissions from non-EU aircraft in its scheme for environmental purposes, it (temporarily) modified its policies ('stopping the clock') after making the assessment that a deal was likely forthcoming through the International Civil Aviation Organization.[31]

3.4 The UN Climate Regime and Its Others: Towards Orchestration

Last but not least, the findings of this book are of relevance for the development of global climate governance more broadly. More specifically, they offer insights into how the UNFCCC can relate to 'its others'.[32]

The negotiations on the future climate regime that started in 2005 have forced policymakers and observers to think about the different roles that the UN climate regime could play. After more than 20 years of international negotiations, it is becoming increasingly apparent that the UNFCCC cannot win the battle against climate change on its own. As I pointed out in Chapter 2, to avoid climate impacts that would most likely be perceived as dangerous by many countries and individuals, a drastic cut in greenhouse gas emissions is not only necessary, but also needs to happen urgently. However, UNFCCC parties have yet to demonstrate their ability, in aggregate, to adopt and implement emission reduction commitments commensurate with the scale and urgency of the climate problem. The reasons for this are varied, but perhaps one of the most important reasons is that the international climate negotiations, like many other debates in world politics, reflect rather than drive national politics, and that mobilizing national constituencies is necessary to drive a global transition towards a low-carbon economy.[33] The ambition level of the UNFCCC and the goodwill of its parties are constrained if the world's major emitters are unwilling to enhance their ambition domestically. Conversely, true leadership on the part of the world's largest absolute and historical emitters would go a long way towards unlocking some of the distrust that currently characterizes the UNFCCC process.

---

[31] European Commission 2012.
[32] The following draws on Moncel and van Asselt 2012.
[33] E.g., Sprinz and Weiß 2001, 67–68.

If it indeed becomes increasingly clear that, faced with a lack of political will from key parties, the UN climate regime can no longer play the role of regulator (i.e., prescribing the emission reductions required for each country *à la* Kyoto) it may instead play an important part as 'orchestrator'.[34] This means, for instance, that through its monitoring and reporting function, it could keep track of the variety of governance arrangements outside the climate regime, and assess whether adding up the efforts of these initiatives is in line with common objectives, such as keeping temperature increases below 2°C (or sticking as close to that as physically possible). Orchestration could also seek to reduce inefficiencies by avoiding duplication of efforts, and decrease the risk of double counting mitigation actions. For example, through the establishment of common guidelines and accounting frameworks, the UNFCCC could ensure the compatibility of existing and emerging emissions trading systems. Similarly, through coordination with minilateral clean technology agreements outside of the UNFCCC, links could be established between funding mechanisms under the climate regime (notably the newly established Green Climate Fund) and technology transfer projects on the ground. By doing so, the UNFCCC could improve coherence in the fragmented climate governance landscape. More generally, the activities described here underscore the various roles international law could play in addition to setting rules and promoting compliance with them.[35]

This potential role of the UNFCCC has come to the fore in the ADP negotiations (see Chapter 2) under the guise of so-called 'international cooperative initiatives',[36] which could help close the pre-2020 mitigation ambition gap.[37] The increasing acknowledgment of this potential contribution has led to calls for 'tracking' and 'recognition' of the contribution of these initiatives.[38] The rationale behind such calls is obvious: it would help assess to what extent initiatives outside the UNFCCC help contribute to climate change mitigation objectives. Furthermore, it could reduce suspicions that such initiatives are meant to distract from the UNFCCC, or that they are merely 'greenwashing' the image of laggards.

---

34 Cf. Abbott and Snidal 2010; Abbott *et al.* 2012b.

35 Cf. Howse and Teitel 2010.

36 These initiatives include a variety of intergovernmental and transnational climate-related initiatives, ranging from the C40 cities initiative to the Montreal Protocol. For a list, see http://unfccc.int/meetings/bonn_jun_2013/items/7655.php (accessed 11 December 2013). See also UNFCCC 2013.

37 UNEP 2013.

38 Hale 2013.

Still, achieving greater coordination in global climate governance and assuming a central role for the UNFCCC in promoting this is not without challenges. One key question that is likely to emerge is where the authority to allocate responsibility for action to one institution over another lies, an issue that plagues interaction management more generally, as noted in the previous section. There is no straightforward answer to this question, and preferred solutions will likely differ from institution to institution. Moreover, in some cases orchestration will be challenging because parties intentionally sought to take climate-related issues elsewhere, as was illustrated by the case of the APP. In other words, the feasibility of orchestration is undermined if countries strategically pursue forum-shopping or forum-creating strategies. Lastly, institutions with similar, partially overlapping mandates may sometimes be more inclined to compete for authority, recognition and financial resources than to delegate power in the broader pursuit of overarching goals (although evidence of such 'turf wars' is anecdotal at best). So while widespread orchestration thus faces important obstacles and will likely take time, one can begin with the 'low-hanging fruit' and make use of informal channels for coordination, and gradually aspire to build a more elaborate and comprehensive network of institutions in the area of global climate governance, with a nodal role for the UNFCCC.

4. THEORETICAL REFLECTIONS AND AREAS FOR FUTURE RESEARCH

This book has sought to contribute to the rapidly expanding body of literature on fragmentation and regime interactions by applying these concepts in the analysis of global climate governance. By doing so, various larger theoretical themes emerged that go well beyond the issue area of climate change. In this final section, I will reflect on these larger themes and identify areas for future research.

4.1 The Politics of Regime Interactions

One of the things that I hope to have made clear throughout this book is that fragmentation and regime interactions are not abstract phenomena appearing out of the blue. While fragmentation can in part be explained by the passing of time, the increasing number of states and other actors, or growing interdependence in an era of globalization,[39] the phenomenon is also a reflection at the global level of clashing social rationalities.[40] One should not assume that these values and norms are simply limited to one regime with, for instance, international economic institutions neatly reflecting values of economic liberalism, and international environmental regimes embodying progressive values related to environmental protection.[41] Instead, different regimes have 'provided a site of competition for a broadly opposed set of goals and interests, *each* competing on *each* terrain'.[42] In other words, values are being contested in the politics of regimes as well as in the politics of regime interactions.

This argument has been made forcefully in the context of the international trade regime by Andrew Lang, who provides an account of how 'political struggles and debates about the social consequences of the global trading system came to be reconstituted as debates about the incoherence of the international legal order, and about the relationship between the trade and human rights regimes'.[43] In doing so, he shows that the 'trade and …' tensions highlighted by scholars in the late 1990s/early 2000s[44] are not merely the result of how regimes relate to each other, but also of how a regime itself evolves. In the case of the trade regime, he argues that 'functional specialization of the trade regime … produced a partiality of vision or bias in the way that it operated',[45] which in turn led to the externalization of environmental, social and cultural issues (i.e., these issues were deemed to be within the domain of other regimes).

Lang's argument can also be explored for regime interactions in global climate governance. Perhaps the clearest example in this book is the climate and trade overlap. At first blush, this seems to be a case where climate protection is being pitched against free trade. Indeed, the issue of unilateral trade measures has been a source of heated debates in both the climate and trade communities. However, I questioned in Chapter 8 whether this dichotomy of trade versus climate should be seen as the real source of those tensions, and argued that there is an alternative explanation that the tension can be traced back to a deeper political divide on how to allocate responsibilities for climate change mitigation: who does

---

39 Raustiala 2013.
40 Cf. Fischer-Lescano and Teubner 2004; see also Krisch 2010.
41 Zelli *et al.* 2012.
42 S. Humphreys 2012, 184.
43 Lang 2011, 127.
44 E.g., Trachtman 2002.
45 Lang 2011, 128.

what and by when? As observers will know, this has been one of the most contentious issues in international climate politics.[46] However, unilateral trade measures adopted by developed countries against (developing) countries that do not take 'comparable action' on climate change can equally be framed as an attempt by one actor to impose domestic standards on other actors. In this case, the standard concerns a certain interpretation of what is an equitable allocation of responsibilities.[47]

There are further examples beyond the scope of this study. A well-known instance is the struggle over the precautionary principle (particularly between thc EU and the US), the status and interpretation of which has been simultaneously contested in different regimes, including the WTO and the Cartagena Protocol on Biosafety.[48] Or to use another famous example, the appropriate level of protection of intellectual property rights is being contested in a variety of regimes.[49] In other words, the same political struggles reappear in several regimes.

However, studies of regime interactions do not always sufficiently engage with the regime-specific political contestations that underpin regime interactions, and tend to view the interacting regimes largely as static 'billiard balls', to use a metaphor previously applied to states.[50] But in the same way that states are not unitary actors, the billiard ball analogy is also not appropriate for regimes, which are shaped by a variety of forces and evolve continuously.[51] The failure to engage with internal politics forms the basis for Koskenniemi's critique of 'managerial' approaches to regime interactions.[52] His argument is valid, and not only for reasons of sound scholarship. 'Solutions' for regime interactions (e.g., permitting or prohibiting climate-related trade measures) that do not engage with the underlying causes for those interactions are unlikely to go far in the real world.

While thus deploring depoliticized approaches to regime interactions, Koskenniemi nevertheless remarks that 'vocabularies that seek to transcend the horizontal juxtaposition between equally partial regimes' can be valuable '[t]o the extent these efforts end up highlighting the political nature of the operation of functional regimes and point to values such as

---

46 E.g., Rajamani 2006.
47 Brewster 2011.
48 Zelli *et al.* 2012, 188.
49 Helfer 2004; Helfer 2009.
50 Wolfers 1951, 40.
51 Cf. Koskenniemi 2007, 27.
52 Koskenniemi 2012.

democracy, transparency, accountability, inclusion and fairness'.[53] In other words, interaction management needs to engage with not only the 'external' politics of regime interactions but also with the 'internal' politics of regimes. Chapter 9 suggests that this is possible. In particular, the chapter reinforced Margaret Young's call for accountable regime interaction 'to ensure ongoing and open contestation between stakeholders rather than entrenched ideals'.[54] Put another way, interacting regimes should not take norms from other regimes as a given, but examine the political process by which these norms were created.

The preceding discussion draws attention to an important area for further research: how do specific actors (including states, but also treaty bodies and other non-state actors) influence regime interactions and their management and, conversely, how do regime interactions influence the strategies of these actors? These lines of inquiry can build on existing research. For instance, by examining the notions of forum shopping,[55] regime shifting[56] and strategic inconsistency,[57] scholars have started exploring states' strategies in a fragmented governance landscape. Similarly, the concepts of bandwagoning[58] and orchestration[59] (see above) denote possible strategies that can be employed by international organizations and treaty bodies to influence regime interactions. There is more to be explored here, however. For instance, deconstructing regime interactions by examining the political struggles within each of the interacting regimes and linking the positions of key actors in those struggles could help explain how and why regime interactions emerge in specific cases.[60] Furthermore, there has only been scant attention to the role of non-state actors, such as non-governmental organizations, the private sector, or public–private partnerships, in shaping regime interactions and interaction management. In particular, the strategies these actors employ when they engage in autonomous interaction management could be further explored.

53 Koskenniemi 2012, 320–321.
54 M.A. Young 2011b, 281.
55 Benvenisti and Downs 2007.
56 Helfer 2004; Helfer 2009.
57 Raustiala and Victor 2004.
58 Jinnah and Muñoz Cabré 2011.
59 Abbott and Snidal 2010; Abbott *et al.* 2012b.
60 Zelli *et al.* 2012.

4.2 From Regime Interactions to Institutional Complexity

Although the focus on regime interactions is an important step away from viewing (and analysing) individual regimes in isolation, it cannot capture the functioning of a variety of institutions in a specific issue area or improve the overall coherence of international law and governance.

Over the last few years, various concepts have been introduced to address this gap (see Chapter 4). Together with colleagues, I have suggested the notion of 'global governance architectures', which can be defined as 'the overarching system of public and private institutions that are valid or active in a given issue area of world politics'.[61] The notion of 'regime complex', coined by Kal Raustiala and David Victor,[62] similarly points to a level of analysis superseding a focus on individual regimes. Oberthür and Stokke instead adopt the term 'institutional complex'.[63] Like Biermann *et al.*, they focus on institutions rather than regimes, allowing for the inclusion of non-state actor initiatives, often overlooked in the regime complexity literature.[64]

These various concepts underline that it is useful to broaden the research agenda from regimes to the functioning of overarching governance structures. Broadly speaking, there are four sets of research questions that need to be tackled under this agenda.[65] First, there are questions regarding *mapping* institutional complexity. A starting point for any such research endeavour would be to define the boundaries of an institutional complex. How one defines a specific issue area will generally be strongly correlated to the assessment of how complex or fragmented that issue area is. For instance, the issue area of global forest governance can be seen as more complex than global REDD governance; and global climate governance will likely be viewed as more complex than the governance of climate change-induced migration. Mapping is therefore a sensible first step, as it can shed light on the boundaries of an institutional complex (even though these boundaries will still be mainly defined by the researchers), clarify which actors and institutions are involved and how they relate to each other, and highlight gaps and overlaps in governance. By using similar mapping criteria, it may then also be possible to compare different institutional complexes.

---

61 Biermann *et al.* 2009, 15.
62 Raustiala and Victor 2004.
63 Oberthür and Stokke 2011b; Oberthür and Stokke 2012.
64 Cf. Abbott 2012, 574–575.
65 The following draws on Zelli and van Asselt 2013.

A second set of research questions focuses on the *causes* of institutional complexity. This book has only scratched the surface of the causes of institutional complexity in global climate governance by indicating that some regime interactions may have been intentional. However, in a way similar to the discussion in the previous section on the politics of regime interactions, the roles and strategies of state and non-state actors in institutional complexes deserve closer scrutiny. Various questions emerge in this regard. Why do some actors create new regimes or turn to other existing regimes in an already dense institutional environment? Can this behaviour be explained by using existing theories from international relations (for instance, regime theory with its focus on power, knowledge and interests[66])? Is there a correlation between the level of policy coherence at the national level and the level of institutional complexity at the international level? Can institutional complexity be related to the structure of the problem at hand or, put differently, does an institutional complex provide a better 'fit' with the socio-ecological dynamics underlying a particular problem?[67]

The third set of research questions extends the questions on *consequences* tackled in this book to institutional complexes. While closely related to the study of the consequences of regime interactions, the focus on institutional complexes raises new questions as well. A key question in this regard is: how does institutional complexity affect the implementation of regimes at the national and sub-national level? This question is particularly pertinent for developing countries with fewer resources at their disposal to effectively and simultaneously implement a variety of international agreements. A related issue is how institutional complexity affects different actors in terms of their ability to navigate complexity. For instance, does complexity benefit powerful countries at the expense of weaker countries, as has been suggested by several authors?[68] More generally, who wins and who loses? The role of non-state actors also merits further analysis; for instance, are some non-governmental organizations better able to benefit from the existence of a variety of international regimes than others?[69] In addition, by analysing more than two regimes, it may also be possible to test theories that emphasize the

66 Zelli 2009; see also Abbott *et al.* 2012a.
67 Cf. O.R. Young 2002.
68 Benvenisti and Downs 2007; Drezner 2009.
69 Orsini 2013.

purported benefits of diversity in global governance, including theories related to experimentation,[70] polycentricity,[71] and redundancy.[72]

The last set of research questions concerns another issue that has been central to this book and deals with the *responses* to institutional complexity. The questions are similar to the ones addressed in this study on the management of regime interactions: What are the options for managing institutional complexity? Which actors are involved in this? Under which conditions could it work? And when can the management of institutional complexity be viewed as successful? Finally, further research could shed light on the ways in which efforts to integrate policies at the national (and sub-national) level could strengthen the coherence of international law and governance. For instance, could increased coordination between the government officials responsible for the implementation of the climate treaties and the government officials responsible for a country's trade policies benefit the broader relationship between the climate and trade regimes? Vice versa, it could be examined whether institutional coordination at the international level could strengthen coherence in the implementation phase.

Research on institutional complexity can form an important complement to the existing research on regime interactions. The latter will remain important, however, as the causes and consequences of, and responses to, institutional complexity will in large part depend on the causes and consequences of, and responses to, regime interactions in the particular issue area. In the same way that studies of regime interactions should not disregard earlier studies of regimes, studies on institutional complexity should not ignore the findings of earlier studies on regime interactions.

4.3 Fragmented Communities? International Law and International Relations

On a final note, I firmly believe that there is room for stronger collaboration between international law and international relations scholars on the subjects of fragmentation and regime interaction.

It is true that developing an interdisciplinary research agenda can be challenging. However, in some respects the lack of awareness of debates in other disciplines is truly surprising. The phenomenon of overlapping international legal instruments, their consequences, and the options for

70 Cf. Hoffmann 2011.
71 Cf. E. Ostrom 2010.
72 Cf. Low *et al.* 2003; P. Haas 2004, 3.

dealing with them have been observed and discussed by international lawyers since the 1950s,[73] yet this literature hardly featured in (early) contributions by international relations scholars on institutional interplay.[74] Perhaps this can be explained by a disinterest in the descriptive/prescriptive approach adopted by lawyers up to that point, with the key questions being about proving the existence of a conflict of norms and determining which norms should prevail. However, it can also be related to the more fundamental scepticism about the quality of international legal theory alluded to by Jeffrey Dunoff and Mark Pollack:

> To the extent that [international relations] scholars consider international legal theory at all, it is primarily to dismiss it as unhelpful, either because it is unduly narrow in its focus upon the language of international legal instruments; politically naïve in devoting substantial attention to unenforceable legal rules but failing adequately to account for power; or methodologically suspect, as legal writings are often normative (urging reform of legal rules or institutions) but rarely positivist (generating and empirically testing causal claims about the world.[75]

Conversely, international law scholarship up to the early 2000s rarely engaged in questions of what caused a specific interaction, and engaged insufficiently with the politics underlying regime interactions. Moreover, the ILC's report on the fragmentation of international law, as well as much of the (extensive) body of literature, almost completely bypasses the work on institutional interactions and interlinkages carried out by international relations scholars since the 1990s (and arguably before that[76]). This can perhaps be explained by an implicit preference for legal techniques in interaction management,[77] obviating the need to engage in discussions of how and why certain types of regime interactions take place. Yet, as I have shown, regime interactions are about more than just litigation or the application of legal techniques. Likewise, the explanation may lie in the scepticism of some international lawyers regarding the discipline of international relations.[78]

---

[73] E.g., Jenks 1953.

[74] E.g., O.R. Young 1996; O.R. Young 2002; Stokke 2001; Oberthür and Gehring 2006.

[75] Dunoff and Pollack 2012, 1–2.

[76] Zelli 2009, 5.

[77] S. Humphreys 2012, 187.

[78] E.g., Koskenniemi 2000; Koskenniemi 2009; Klabbers 2005.

The attitudes described here seem to be changing, however. The flourishing research agendas on fragmentation, regime/institutional complexity, and regime interactions show that there is an increasing awareness on both sides.[79] This study supports the argument that this is a welcome development, by pointing to several lessons that the disciplines can learn from each other.

First, this book shows that from the perspective of international law, it matters whether a conflict can be classified as a normative conflict or not. A true conflict in the legal sense gives rise to a variety of legal problems, as is evidenced by the discussion in Chapters 3 and 4. While some of the fears expressed by international lawyers about the fragmentation of international law may have been exaggerated (or inspired by personal views), the fact remains that in some cases legal actors are faced with norms that are incompatible with each other. Furthermore, the attention paid by international lawyers to the use of legal techniques is not merely an academic exercise: international adjudicators often face challenging situations where they have to harmonize norms from different regimes.

Conversely, from the perspective of international relations, it is clear that regime interactions lead to consequences that are not always anticipated or discussed by international lawyers. I have argued that the synergies that regime interactions may produce have by and large been ignored. In addition, broader tensions between regimes that cannot be characterized as normative conflicts have received less attention than the situations in which norms clashed or threatened to clash.

Second, this book shows how the international law literature has provided important insights into soft–hard law interactions. This started with the early literature in the 1980s/1990s, which showed how soft law could complement hard law, for instance by paving the way for, or elaborating on, hard law. As I discussed in Chapter 4, more recent contributions to the legal literature show how soft and hard law can also act as antagonists.[80] This is an important observation, as it draws attention to the conditions under which soft and hard law may complement or undermine each other. These conditions may relate to the fragmented governance landscape as well as the problem structure at hand, and require further exploration.[81]

Third, critical international legal scholars underscore that 'managing' regime interactions is not and *should not* be devoid of politics. As I

---

[79] See, e.g., Raustiala 2013.

[80] Shaffer and Pollack 2010; Shaffer and Pollack 2011; Pollack and Shaffer 2012.

[81] See, for an initial discussion, Shaffer and Pollack 2013.

argued above, I do not think that the risk of 'managerialism' is equally high in all cases of regime interaction and interaction management. However, it is important to remember that regime interactions do not emerge spontaneously and that the management of interactions may have important consequences for certain actors. By showing how regime interactions are intrinsically related to the internal politics of regimes, international lawyers have drawn attention to underlying challenges to interaction management.

Nevertheless, the two disciplines still have much more to learn from each other in the further analysis of fragmentation and complexity in international law and governance.

First, ongoing discussions in international law on global or international constitutionalization point to efforts on the part of international lawyers to either describe whether there is a global constitution[82] or to (normatively) argue that there should (not) be one.[83] The broader debate on constitutionalization among international lawyers may help better define the 'redemptive narrative' for reconciling different regimes[84] by providing guidance on whether any overarching yardsticks for interaction management could be identified and, if so, how these could possibly be operationalized for specific regime interactions.

Third, there is room for further inquiry into how soft and hard law coexist in dense institutional environments. Beyond the previously mentioned typology of interactions between hard and soft law, it is interesting to examine how the advantages and disadvantages of hard and soft law play out in such cases, and how their strengths and weaknesses can be used to ensure they complement each other. For instance, soft law is often associated with increased flexibility and adaptiveness, whereas hard law is associated with predictability and certainty. Is it then possible to combine these virtues? A related strand of research would be to examine why states pursue formal and informal institutions in a fragmented environment.[85]

Lastly, the two disciplines have more to learn from each other when it comes to examining the accountability of non-state actors in addressing regime interactions. In this book, key non-state actors have been the

---

[82] E.g., Fassbender 1998.

[83] See, for an example in the context of international environmental law, Bodansky 2009a.

[84] Dunoff 2012, 155.

[85] This could form an extension of the research agenda on international institutional choice. See Abbott and Snidal 1998; Raustiala 2005; Jupille and Snidal 2006.

bureaucracies involved in interaction management. As Chapter 9 highlighted, their involvement has raised accountability and legitimacy concerns. Largely independently of each other, both disciplines have sought to address these issues. As discussed briefly in Chapter 3 under the projects on 'global administrative law' and the 'exercise of international public authority', international lawyers have sought to explore legal principles and norms that could regulate the activities of non-state actors at the international level. Similarly, international relations scholars have explored how to improve the accountability and legitimacy of earth system governance.[86] The research agendas are clearly related, but could be more strongly integrated, particularly with regard to the accountability of non-state actors involved in managing regime interactions.

In sum, the important questions about the fragmentation of international law and on regime interactions should not only be tackled by international lawyers, nor should they be dealt with solely from the point of view of international relations theory. It is time to build on the strengths of both communities.

---

[86] E.g., Biermann and Gupta 2011.

References

Abbott, Kenneth W. (1989), 'Modern International Relations Theory: A Prospectus for International Lawyers', *Yale Journal of International Law*, 14(2), 335–411.

Abbott, Kenneth W. (2012), 'The Transnational Regime Complex for Climate Change', *Environment & Planning C: Government & Policy*, 30(4), 571–590.

Abbott, Kenneth W. and Duncan Snidal (1998), 'Why States Act through Formal International Organizations', *Journal of Conflict Resolution*, 42(1), 3–32.

Abbott, Kenneth W. and Duncan Snidal (2000), 'Hard and Soft Law in International Governance', *International Organization*, 54(3), 421–456.

Abbott, Kenneth W. and Duncan Snidal (2010), 'International Regulation without International Government: Improving IO Performance through Orchestration', *Review of International Organizations*, 5(3), 315–344.

Abbott, Kenneth W., Robert O. Keohane, Andrew Moravcsik, Anne-Marie Slaughter and Duncan Snidal (2000), 'The Concept of Legalization', *International Organization*, 54(3), 401–419.

Abbott, Kenneth W., Jessica F. Green and Robert O. Keohane (2012a), 'Institutional Density and the Ecology of Public and Private Institutions in World Politics', available at http://www.princeton.edu/politics/about/file-repository/public/AGKPrinceton.pdf (accessed 11 December 2013).

Abbott, Kenneth W., Philipp Genschel, Duncan Snidal and Bernhard Zangl (2012b), 'Orchestration: Global Governance through Intermediaries', available at http://papers.ssrn.com/sol3/papers.cfm?abstract_id=2125452 (accessed 11 December 2013).

Abi-Saab, Georges (1999), 'Fragmentation or Unification: Some Concluding Remarks', *New York University Journal of International Law and Politics*, 31(4), 919–933.

Abraham, Spencer (2004), 'The Bush Administration's Approach to Climate Change', *Science*, 305(5684), 616–617.

Allen, Myles R., David J. Frame, Chris Huntingford, Chris D. Jones, Jason A. Lowe, Malte Meinshausen and Nicolai Meinshausen (2009a), 'Warming Caused by Cumulative Carbon Emissions towards the Trillionth Tonne', *Nature*, 458(7242), 1163–1166.

Allen, Myles, David Frame, Katja Frieler, William Hare, Chris Huntingford, Chris Jones, Reto Knutti, Jason Lowe, Malte Meinshausen, Nicolai Meinshausen and Sarah Raper (2009b), 'The Exit Strategy', *Nature Reports Climate Change*, 3, 56–58.

Alter, Karen J. and Sophie Meunier (2009). 'The Politics of International Regime Complexity', *Perspectives on Politics*, 7(1), 13–24.

Anderson, Kevin and Alice Bows (2011), 'Beyond 'Dangerous' Climate Change: Emission Scenarios for a New World', *Philosophical Transactions of the Royal Society* A, 369, 20–44.

Anton, Donald K. (2012), '"Treaty Congestion" in Contemporary International Environmental Law', in Shawkat Alam, Jahid Hossain Bhuiyan, Tareq M.R. Chowdhury and Erika J. Techera (eds.), *Routledge Handbook of International Environmental Law*, London, UK: Routledge, 651–665.

APP Charter (2006), 'Charter for the Asia-Pacific Partnership on Clean Development and Climate', available at http://www.asiapacificpartnership.org/pdf/resources/charter.pdf (accessed 11 December 2013).

APP Vision Statement (2006), 'Vision Statement of Australia, China, India, Japan, the Republic of Korea, and the United States of America for a New Asia-Pacific Partnership on Clean Development and Climate', available at http://www.asiapacificpartnership.org/pdf/resources/vision.pdf (accessed 11 December 2013).

Assunção, Lucas and ZhongXiang Zhang (2002), *Domestic Climate Policies and the WTO*, United Nations Conference on Trade and Development Discussion Paper No. 164, Geneva: United Nations Conference on Trade and Development.

Aust, Anthony (2007), *Modern Treaty Law and Practice*, Cambridge, UK: Cambridge University Press.

Austin, Kemen, Florence Daviet and Fred Stolle (2010), 'The REDD+ Decision in Cancun', available at http://www.wri.org/stories/2010/12/redd-decision-cancun (accessed 11 December 2013).

Axelrod, Mark (2011), 'Savings Clauses and the "Chilling Effect": Regime Interplay as Constraints on International Governance', in Sebastian Oberthür and Olav Schram Stokke (eds.), *Managing Institutional Complexity: Regime Interplay and Global Environmental Change*, Cambridge, MA: The MIT Press, 87–114.

Azar, Christian and Henning Rodhe (1997), 'Targets for Stabilization of Atmospheric CO_2', *Science*, 276(5320), 1818–1819.

Bäckstrand, Karin (2008), 'Accountability of Networked Climate Governance: The Rise of Transnational Climate Partnerships', *Global Environmental Politics*, 8(3), 74–102.

Bäckstrand, Karin and Eva Lövbrand (2006), 'Planting Trees to Mitigate Climate Change: Contested Discourses of Ecological Modernization, Green Governmentality, and Civic Environmentalism', *Global Environmental Politics*, 6(1), 50–75.

Bailey, Zoya E. (2002), 'The Sink That Sank The Hague: A Comment on the Kyoto Protocol', *Temple International and Comparative Law Journal*, 16(1), 106–121.

Bakker, Stefan, Constanze Haug, Harro van Asselt, Raouf Saïdi and Joyeeta Gupta (2011), 'The Future of the CDM: Same Same, but Differentiated?', *Climate Policy*, 11(1), 752–767.

Barrett, Scott (2003), *Environment and Statecraft: The Strategy of Environmental Treaty-Making*, Oxford: Oxford University Press.

Barrett, Scott (2010), 'Climate Change and International Trade: Lessons on their Linkage from International Environmental Agreements', available at http://www.wto.org/english/res_e/reser_e/climate_jun10_e/background_paper6_e.pdf (accessed 11 December 2013).

Barrett, Scott and Robert N. Stavins (2003), 'Increasing Participation and Compliance in International Climate Change Agreements', *International Environmental Agreements: Politics, Law and Economics*, 3(4), 349–376.

Barton, John H. (2007), *Intellectual Property and Access to Clean Energy Technologies in Developing Countries: An Analysis of Solar Photovoltaic, Biofuel and Wind Technologies*, Geneva: International Centre for Trade and Sustainable Development.

Bausch, Camilla and Michael Mehling (2006), '"Alive and Kicking": The First Meeting of the Parties to the Kyoto Protocol', *Review of European Community and International Environmental Law*, 15(2), 193–201.

Bausch, Camilla and Michael Mehling (2011), *Addressing the Challenge of Global Climate Mitigation – An Assessment of Existing Venues and Institutions*, Berlin: Friedrich-Ebert Stiftung.

Baxter, Richard R. (1980), 'International Law in "Her Infinite Variety"', *International and Comparative Law Quarterly*, 29(4), 549–566.

BBC (2011), 'Canada to Withdraw from Kyoto Protocol', available at http://www.bbc.co.uk/news/world-us-canada-16151310 (accessed 11 December 2013).

BBC (2013), 'Carbon Dioxide Passes Symbolic Mark', available at http://www.bbc.co.uk/news/science-environment-22486153 (accessed 11 December 2013).

Beck, Robert J. (2009), 'International Law and International Relations Scholarship', in James David Armstrong (ed.), *Routledge Handbook of International Law*, Abingdon, UK: Routledge, 13–43.

Benioff, Ron, Heleen de Coninck, Subash Dhar, Ulrich Hansen, Joyce McLaren and Jyoti Painuly (2010), *Strengthening Clean Energy Technology Cooperation under the UNFCCC: Steps toward Implementation*, Golden, CO: National Renewable Energy Laboratory.

Benvenisti, Eyal and George W. Downs (2007), 'The Empire's New Clothes: Political Economy and the Fragmentation of International Law', *Stanford Law Review*, 60(2), 595–632.

Berkhout, Frans (2005), 'Rationales for Adaptation in EU Climate Change Policies', *Climate Policy*, 5(3), 377–391.

Berman, Paul Schiff (2007), 'Global Legal Pluralism', *Southern California Law Review*, 80(6), 1155–1238.

Bernasconi-Osterwalder, Nathalie and Johannes Norpoth (2009), *Is World Trade Law a Barrier to Saving Our Climate? Questions and Answers*, Washington, DC: Center for International Environmental Law.

Bernstein, Steven and Jutta Brunnée (2011), 'Options for Broader Reform of the Institutional Framework for Sustainable Development (IFSD): Structural, Legal, and Financial Aspects', available at http://www.uncsd2012.org/index.php?page=view&type=400&nr=211&menu=45 (accessed 11 December 2013).

Bernstein, Steven and Maria Ivanova (2007), 'Institutional Fragmentation and Normative Compromise in Global Environmental Governance: What Prospects for Re-embedding?', in Steven Bernstein and Louis W. Pauly (eds.), *Global Liberalism and Political Order: Towards a New Grand Compromise?*, Albany, NY: State University of New York Press, 161–185.

Bernstein, Steven, Michele Betsill, Matthew Hoffmann and Matthew Paterson (2010), 'A Tale of Two Copenhagens: Carbon Markets and Climate Governance', *Millennium – Journal of International Studies*, 39(1), 161–173.

Betsill, Michele and Harriet Bulkeley (2004), 'Transnational Networks and Global Environmental Governance: The Cities for Climate Protection Program', *International Studies Quarterly*, 48(2), 471–493.

Bhagwati, Jagdish and Petros C. Mavroidis (2007), 'Is Action against US Exports for Failure to Sign Kyoto Protocol WTO-legal?', *World Trade Review*, 6(2), 299–310.

Biermann, Frank (2005), 'The Rationale for a World Environment Organization', in Frank Biermann and Steffen Bauer (eds.), *A World Environment Organization: Solution or Threat for Effective International Environmental Governance?*, Aldershot, UK: Ashgate, 117–144.

Biermann, Frank (2007), '"Earth System Governance" as a Crosscutting Theme of Global Change Research', *Global Environmental Change*, 17(3–4), 326–337.

Biermann, Frank and Steffen Bauer (eds.) (2005), *A World Environment Organization: Solution or Threat for Effective International Environmental Governance?*, Aldershot, UK: Ashgate.

Biermann, Frank and Rainer Brohm (2005), 'Implementing the Kyoto Protocol without the United States: The Strategic Role of Energy Tax Adjustments at the Border', *Climate Policy*, 4(3), 289–302.

Biermann, Frank and Aarti Gupta (2011), 'Accountability and Legitimacy in Earth System Governance: A Research Framework', *Ecological Economics*, 70(11), 1856–1864.

Biermann, Frank and Bernd Siebenhüner (2009a), 'The Role and Relevance of International Bureaucracies: Setting the Stage', in Frank Biermann and Bernd Siebenhüner (eds.), *Managers of Global Change: The Influence of International Environmental Bureaucracies*, Cambridge, MA: The MIT Press, 1–14.

Biermann, Frank and Bernd Siebenhüner (eds.) (2009b), *Managers of Global Change: The Influence of International Environmental Bureaucracies*, Cambridge, MA: The MIT Press.

Biermann, Frank, Philipp Pattberg, Harro van Asselt and Fariborz Zelli (2009), 'The Fragmentation of Global Governance Architectures: A Framework for Analysis', *Global Environmental Politics*, 9(4), 14–40.

Bigdeli, Sadeq Z. (2008), 'Will the "Friends of Climate" Emerge in the WTO? The Prospects of Applying the "Fisheries Subsidies" Model to Energy Subsidies', *Carbon and Climate Law Review*, 2(1), 78–88.

Birnie, Patricia and Alan Boyle (2002), *International Law and the Environment*, 2nd edition, Oxford: Oxford University Press.

Bodansky, Daniel M. (1993), 'The United Nations Framework Convention on Climate Change: A Commentary', *Yale Journal of International Law*, 18(2), 451–558.

Bodansky, Daniel (2006), 'Does One Need to Be an International Lawyer to Be an International Environmental Lawyer?', *Proceedings of the American Society of International Law*, 100, 303–307.

Bodansky, Daniel (2009a), 'Is There an International Environmental Constitution?', *Indiana Journal of Global Legal Studies*, 16(2), 565–584.

Bodansky, Daniel (2009b), *The Art and Craft of International Environmental Law*, Cambridge, MA: Harvard University Press.

Bodansky, Daniel (2010), 'The Copenhagen Climate Change Conference: A Postmortem', *American Journal of International Law*, 104(2), 230–240.

Bodansky, Daniel, Jutta Brunnée and Ellen Hey (2007), 'International Environmental Law: Mapping the Field', in Daniel Bodansky, Jutta Brunnée and Ellen Hey (eds.), *The Oxford Handbook of International Environmental Law*, Oxford: Oxford University Press, 1–28.

Bordoff, Jason E. (2009), 'International Trade Law and the Economics of Climate Policy: Evaluating the Legality and Effectiveness of Proposals to Address Competitiveness and Leakage Concerns', *Brookings Trade Forum*, 2008/2009, 35–59.

Borgen, Christoph J. (2005), 'Resolving Treaty Conflicts', *George Washington International Law Review*, 37(3), 573–648.

Boyle, Alan (1999), 'Some Reflections on the Relationship of Treaties and Soft Law', *International and Comparative Law Quarterly*, 48(4), 901–913.

Breitmeier, Helmut, Arild Underdal and Oran R. Young (2011), 'The Effectiveness of International Environmental Regimes: Comparing and Contrasting Findings from Quantitative Research', *International Studies Review*, 13(4), 579–605.

Brewer, Thomas L. (2003), 'The Trade Regime and the Climate Regime: Institutional Evolution and Adaptation', *Climate Policy*, 3(4), 329–341.

Brewer, Thomas L. (2004), 'The WTO and the Kyoto Protocol: Interaction Issues', *Climate Policy*, 4(1), 3–12.

Brewer, Thomas L. (2008), 'Climate Change Technology Transfer: A New Paradigm and Policy Agenda', *Climate Policy*, 8(5), 516–526.

Brewer, Thomas L. (2009), 'Technology Transfers and Climate Change: International Flows, Barriers, and Frameworks', *Brookings Trade Forum*, 2008/2009, 93–119.

Brewster, Rachel (2011), 'The Form and Substance of Carbon Tariffs', paper prepared for the Conference on International Relations and Climate Change, Princeton University, 18–19 February 2011.

Broadbent, Jeffrey P. (2002), 'From Heat to Light?: Japan's Changing Response to Global Warming', in John D. Montgomery and Nathan Glazer (eds.), *Sovereignty under Challenge*, New Brunswick, NJ: Transaction Publishers, 109–142.

Broude, Tomer (2008), 'Principles of Normative Integration and the Allocation of International Authority: The WTO, the Vienna Convention on the Law of Treaties, and the Rio Declaration', *Loyola University Chicago International Law Review*, 6(1), 173–207.

Brown, David, Frances Seymour and Leo Peskett (2008), 'How Do We Achieve REDD Co-benefits and Avoid Doing Harm?', in Arild Angelsen (ed.), *Moving Ahead with REDD: Issues, Options and Implications*, Bogor: Center for International Forestry Research, 107–118.

Brown Weiss, Edith (1993), 'International Environmental Law: Contemporary Issues and the Emergence of a New Order', *Georgetown Law Journal*, 81(3), 675–710.

Brown Weiss, Edith and Harold K. Jacobson (eds.) (1998), *Engaging Countries: Strengthening Compliance with International Environmental Accords*, Cambridge, MA: The MIT Press.

Brunnée, Jutta (2002), 'COPing with Consent: Law Making under Multilateral Environmental Agreements', *Leiden Journal of International Law*, 15(1), 1–52.

Brunnée, Jutta (2005), 'Reweaving the Fabric of International Law. Patterns of Consent in Environmental Framework Agreements', in Rüdiger Wolfrum and Volker Röben (eds.), *Developments of International Law in Treaty-Making*, Berlin: Springer, 101–126.

Buck, Matthias and Roda Verheyen (2001), *International Trade Law and Climate Change – A Positive Way Forward*, Bonn: Friedrich Ebert Stiftung.

Burke-White, William (2004), 'International Legal Pluralism', *Michigan Journal of International Law*, 25(4), 963–979.

Burns, William C.G. (2009), 'Belts and Suspenders: The World Heritage Convention's Role in Confronting Climate Change', *Review of European Community and International Environmental Law*, 18(2), 148–163.

Busch, Per-Olof (2009), 'The Climate Secretariat: Making a Living in a Straitjacket', in Frank Biermann and Bernd Siebenhüner (eds.), *Managers of Global Change: The Influence of International Environmental Bureaucracies*, Cambridge, MA: The MIT Press, 245–264.

Caddell, Richard (2012), 'The Integration of Multilateral Environmental Agreements: Lessons from the Biodiversity-Related Conventions', *Yearbook of International Environmental Law*, 22, 1–39.

Cameron, Edward and Marc Limon (2012), 'Restoring the Climate by Realizing Rights: The Role of the International Human Rights System', *Review of European Community and International Environmental Law*, 21(3), 204–219.

Canada (2007), 'Strong Leadership. A Better Canada. Speech from the Throne', available at http://www.speech.gc.ca/eng/full-speech (accessed 11 December 2013).

Canada (2008), 'Country Statement – Canada', available at http://www.asiapacificpartnership.org/pdf/vancouver/CountryStatement_Canada.pdf (accessed 11 December 2013).

Carlarne, Cinnamon Piñon (2008), 'Good Climate Governance: Only a Fragmented System of International Law Away?', *Law & Policy*, 30(4), 450–480.

Carlarne, Cinnamon Piñon (2010), *Climate Change Law and Policy: EU and US Approaches*, Oxford: Oxford University Press.

Carlino, Hernan, Martina Chidiak and Daniel Perczyk (2007), 'Architecture for Global Climate Technology Governance: Options and Challenges', paper prepared for the Amsterdam Conference on Earth System Governance: Theories and Strategies for Sustainability, Amsterdam, The Netherlands, 24–26 May 2007.

Cash, David W., W. Neil Adger, Fikret Berkes, Po Garden, Louis Lebel, Per Olsson, Lowell Pritchard and Oran Young (2006), 'Scale and Cross-Scale Dynamics: Governance and Information in a Multilevel World', *Ecology and Society*, 11(2), 8.

Cassimatis, Anthony (2007), 'International Humanitarian Law, International Human Rights Law, and Fragmentation of International Law', *International and Comparative Law Quarterly*, 56(3), 623–639.

CBD (2000), *Biological Diversity and Climate Change, Including Cooperation with the United Nations Framework Convention on Climate Change, Note by the Executive Secretary*, UN Doc. UNEP/CBD/SBSTTA/6/11 (21 December 2000).

CBD (2001), *Recommendation VI/7, Biological Diversity and Climate Change, Including Cooperation with the United Nations Framework Convention on Climate Change*, UN Doc. UNEP/CBD/COP/6/3 (27 March 2001).

CBD (2004), *Options for Enhanced Cooperation among the Three Rio Conventions*, UN Doc. UNEP/CBD/SBSTTA/10/INF/9 (15 December 2004).

CBD (2007), *In-depth Review of the Expanded Programme of Work on Forest Biological Diversity. Note by the Executive Secretary*, UN Doc. UNEP/CBD/SBSTTA/13/3 (13 November 2007).

CBD (2010a), *Outcomes of the Global Expert Workshop on Biodiversity Benefits of Reducing Emissions from Deforestation and Forest Degradation*, UN Doc. UNEP/CBD/COP/10/INF/20 16 (1 October 2010).

CBD (2010b), *In-Depth Review of the Work on Biodiversity and Climate Change*, UN Doc. UNEP/CBD/SBSTTA/14/6 (10 February 2010).

CBD (2010c), *Proposed Elements for a Joint Work Programme Between the Three Rio Conventions on Biodiversity, Climate Change and Land Degradation, Note by the Executive Secretary*, UN Doc. UNEP/CBD/SBSTTA/14/6/Add.2 (16 February 2010).

CBD (2012a), *Advice on the Application of Relevant REDD+ Safeguards for Biodiversity, and on Possible Indicators and Potential Mechanisms to Assess Impacts of REDD+ Measures on Biodiversity. Note by the Secretary*, UN Doc. UNEP/CBD/SBSTTA/16/9 (6 March 2012).

CBD (2012b), *Proposals on Integrating Biodiversity Considerations into Climate Change-related Activities, Including Addressing Gaps in Knowledge and Information. Note by the Secretary*, UN Doc. UNEP/CBD/SBSTTA/16/8 (21 February 2012).

CBD (2012c), *Submission by the Secretariat of the Convention on Biological Diversity to the Secretariat of the United Nations Framework Convention on Climate Change. Note by the Secretary*, UN Doc. UNEP/CBD/SBSTTA/16/INF/2 (22 February 2012).

Chambers, W. Bradnee (2008), *Interlinkages and the Effectiveness of Multilateral Environmental Agreements*, Tokyo: United Nations University Press.

Chambers, W. Bradnee (ed.) (1998), *Global Climate Governance: Inter-Linkages between the Kyoto Protocol and Other Multilateral Regimes*, Tokyo: United Nations University Press.

Chambers, W. Bradnee (ed.) (2001), *Inter-linkages: The Kyoto Protocol and the International Trade and Investment Regimes*, Tokyo: United Nations University Press.

Charney, Jonathan (1999), 'The Impact on the International Legal System of the Growth of International Courts and Tribunals', *New York University Journal of International Law and Politics*, 31(4), 697–708.

Charnovitz, Steve (2002), 'The Law of Environmental "PPMs" in the WTO: Debunking the Myth of Illegality', *Yale Journal of International Law*, 27(1), 59–110.

Charnovitz, Steve (2003), 'Trade and Climate: Potential Conflicts and Synergies', in Elliot Diringer (ed.), *Beyond Kyoto: Advancing the International Effort against Climate Change*, Washington, DC: Pew Center for Global Climate Change, 141–170.

Charnovitz, Steve (2007), 'The WTO's Environmental Progress', *Journal of International Economic Law*, 10(3), 685–706.

Charnovitz, Steve (2010a), 'Book Review: Trade and Climate Change: A Report by the United Nations Environment Programme and the World Trade Organization', *World Trade Review*, 9(1), 273–288.

Charnovitz, Steve (2010b), 'Trade and Climate Change: Reviewing Carbon Charges and Free Allowances under Environmental Law and Principles', *ILSA Journal of International and Comparative Law*, 16(2), 395–412.

Chayes, Abram and Antonia Handler Chayes (1995), *The New Sovereignty: Compliance with International Regulatory Agreements*, Cambridge, MA: Harvard University Press.

Childs Staley, Britt and Casey Freeman (2009), *Tick Tech Tick Tech: Coming to Agreement on Technology in the Countdown to Copenhagen*, Washington, DC: World Resources Institute.

Chinese Ministry of Foreign Affairs (2006), 'China–EU Partnership on Climate Change Rolling Work Plan', available at http://www.mfa.gov.cn/eng/wjb/zzjg/tyfls/tfsxw/t283051.htm (accessed 11 December 2013).

Chinkin, Christine (2000), 'Normative Development in the International Legal System', in Dinah Shelton (ed.), *Commitment and Compliance: The Role of Non-Binding Norms in the International Legal System*, Oxford: Oxford University Press, 21–42.

Christoff, Peter and Robyn Eckersley (2007), 'The Kyoto Protocol and the Asia Pacific Partnership on Clean Development and Climate', in Tim Bonyhady and Peter Christoff (eds.), *Climate Law in Australia*, Sydney: Federation Press, 32–45.

Churchill, Robin R. and Geir Ulfstein (2000), 'Autonomous Institutional Arrangements in Multilateral Environmental Agreements: A Little-Noticed Phenomenon in International Law', *American Journal of International Law*, 94(4), 623–659.

Cohen, Harlan Grant (2012), 'Finding International Law, Part II: Our Fragmenting Legal Community', *New York University Journal of International Law and Politics*, 44(4), 1049–1107.

Cole, Daniel H. (2008), 'Climate Change and Collective Action', *Current Legal Problems*, 61(1), 229–264.

Collaborative Partnership on Forests (2003), 'Policy Document', available at http://www.fao.org/forestry/3721-0d87afb466964d3de594709cce3ab941.pdf (accessed 11 December 2013).

Commission on Global Governance (1995), *Our Global Neighborhood. Report of the Commission on Global Governance*, Oxford: Oxford University Press.

Cosbey, Aaron (2008), *Border Carbon Adjustment*, available at http://www.iisd.org/pdf/2008/cph_trade_climate_border_carbon.pdf (accessed 11 December 2013).

Cosbey, Aaron (2009), 'Achieving Consensus: Multilateral Trade Measures in Post-2012 Scenario', in Benjamin Simmons, Harro van Asselt and Fariborz Zelli (eds.), *Climate and Trade Policies in a Post-2012 World*, Geneva: UNEP, 19–26.

Cosbey, Aaron, Susanne Droege, Carolyn Fischer, Julia Reinaud, John Stephenson, Lutz Weischer and Peter Wooders (2012), *A Guide for the Concerned: Guidance on the Elaboration and Implementation of Border Carbon Adjustment*, Stockholm: ENTWINED.

Cossy, Mireille and Gabrielle Marceau (2009), 'Institutional Challenges to Enhance Policy Co-ordination – How WTO Rules Could be Utilised to Meet Climate Objectives?', in Thomas Cottier, Olga Nartova and Sadeq Z. Bigdeli (eds.), *International Trade Regulation and the Mitigation of Climate Change*, Cambridge, UK: Cambridge University Press, 371–394.

Craven, Matthew (2005), 'Unity, Diversity and the Fragmentation of International Law', *The Finnish Yearbook of International Law*, 14, 3–34.

Das, Kasturi (2011), *Technology Transfer under the Clean Development Mechanism: An Empirical Study of 1000 CDM Projects*, The Governance of Clean Development Working Paper 014, Norwich, UK: University of East Anglia.

de Bruyn, Sander, Dagmar Nelissen and Marnix Koopman (2013), *Impact of Recent Developments in the EU ETS on the List of Sectors Deemed to Be Exposed to Carbon Leakage*, Delft: CE Delft.

de Cendra de Larragán, Javier (2006), 'Can Emissions Trading Schemes be Coupled with Border Tax Adjustments?: An Analysis vis-à-vis WTO Law', *Review of European Community and International Environmental Law*, 15(2), 131–145.

de Coninck, Heleen C. (2009), *Technology Rules! Can Technology-Oriented Agreements Help Address Climate Change?*, PhD thesis, VU University Amsterdam.

de Coninck, Heleen, Carolyn Fischer, Richard G. Newell and Takahiro Ueno (2008), 'International Technology-Oriented Agreements to Address Climate Change', *Energy Policy*, 36(1), 335–356.

Dechezleprêtre, Antoine, Matthie Glachant and Yann Ménière (2009), 'Technology Transfer by CDM Projects: A Comparison of Brazil, China, India and Mexico', *Energy Policy*, 37(2), 703–711.

Delimatsis, Panagiotis (2011), 'The Fragmentation of International Trade Law', *Journal of World Trade*, 45(1), 87–116.

Depledge, Joanna (2000), *Tracing the Origins of the Kyoto Protocol: An Article-by-Article Textual History*, UN Doc. FCCC/TP/2000/2 (25 November 2000).

Depledge, Joanna and Farhana Yamin (2009), 'The Global Climate-Change Regime: A Defence', in Dieter Helm and Cameron Hepburn (eds.), *The Economics and Politics of Climate Change*, Oxford: Oxford University Press, 433–453.

Derksen, Samantha (2011), *EU Climate Policies and Developing Country Trade Vulnerability: An Overview of Carbon Leakage-Sensitive Trade Flow*, Geneva: International Centre for Trade and Sustainable Development.

Dingwerth, Klaus and Philipp Pattberg (2006), 'Global Governance as a Perspective on World Politics', *Global Governance*, 12(2), 185–203.

Doelle, Meinhard (2004), 'Linking the Kyoto Protocol and Other Multilateral Environmental Agreements: From Fragmentation to Integration?', *Journal of Environmental Law and Practice*, 14, 75–104.

Douma, Wybe Th. (2006), 'The European Union, Russia and the Kyoto Protocol', in Kurt Deketelaere and Marjan Peeters (eds.), *EU Climate*

Change Policy: The Challenge of New Regulatory Initiatives, Cheltenham, UK and Northampton, MA: Edward Elgar, 51–66.

Drezner, Daniel W. (2009), 'The Power and Peril of International Regime Complexity', *Perspectives on Politics*, 7(1), 65–70.

Dröge, Susanne (2009), *Tackling Leakage in a World of Unequal Carbon Prices. Synthesis Report*, Cambridge, UK: Climate Strategies.

Dunoff, Jeffrey L. (1998), 'Border Patrol at the World Trade Organization', *Yearbook of International Environmental Law*, 9, 20–27.

Dunoff, Jeffrey L. (2012), 'A New Approach to Regime Interaction', in Margaret A. Young (ed.), *Regime Interaction in International Law: Facing Fragmentation*, Cambridge, UK: Cambridge University Press, 136–174.

Dunoff, Jeffrey L. and Mark A. Pollack (2012), 'What Can International Relations Learn from International Law?', available at http://papers.ssrn.com/sol3/papers.cfm?abstract_id=2037299 (accessed 11 December 2013).

Dunoff, Jeffrey L. and Mark A. Pollack (2013a), 'International Law and International Relations: Introducing an Interdisciplinary Dialogue', in Jeffrey L. Dunoff and Mark A. Pollack (eds.), *Interdisciplinary Perspectives on International Law and International Relations: The State of the Art*, Cambridge, UK: Cambridge University Press, 3–32.

Dunoff, Jeffrey L. and Mark A. Pollack (2013b), 'Reviewing Two Decades of IL/IR Scholarship: What We've Learned, What's Next', in Jeffrey L. Dunoff and Mark A. Pollack (eds.), *Interdisciplinary Perspectives on International Law and International Relations: The State of the Art*, Cambridge, UK: Cambridge University Press, 626–662.

Dunoff, Jeffrey L. and Joel P. Trachtman (2009a), 'A Functional Approach to International Constitutionalization', in Jeffrey L. Dunoff and Joel P. Trachtman (eds.), *Ruling the World? Constitutionalism, International Law, and Global Governance*, Cambridge, UK: Cambridge University Press, 3–36.

Dunoff, Jeffrey L. and Joel P. Trachtman (eds.) (2009b), *Ruling the World? Constitutionalism, International Law, and Global Governance*, Cambridge, UK: Cambridge University Press.

Dupuy, Pierre-Marie (1999), 'The Danger of Fragmentation or Unification of the International Legal System and the International Court of Justice', *New York University Journal of International Law and Politics*, 31(4), 791–807.

Ebeling, Johannes and Jan Fehse (2009), *Challenges for a Business Case for High-Biodiversity REDD Projects and Schemes*, Oxford: EcoSecurities.

Eckersley, Robyn (2004), 'The Big Chill: The WTO and Multilateral Environmental Agreements', *Global Environmental Politics*, 4(2), 24–40.

Eckersley, Robyn (2009), 'Understanding the Interplay between the Climate and Trade Regimes', in Benjamin Simmons, Harro van Asselt and Fariborz Zelli (eds.), *Climate and Trade Policies in a Post-2012 World*, Geneva: UNEP, 11–18.

Eckersley, Robyn (2010), 'The Politics of Carbon Leakage and the Fairness of Border Measures', *Ethics & International Affairs*, 24(4), 367–393.

Eckersley, Robyn (2012), 'Moving Forward in the Climate Negotiations: Multilateralism or Minilateralism?', *Global Environmental Politics*, 12(2), 24–42.

Ecosecurities (2010), *The Forest Carbon Offsetting Report 2010*, Oxford: Ecosecurities.

Epps, Tracey and Andrew Green (2010), *Reconciling Trade and Climate: How the WTO Can Help Address Climate Change*, Cheltenham, UK and Northampton, MA: Edward Elgar.

Esty, Daniel C. (1994), *Greening the GATT: Trade, Environment, and the Future*, Washington, DC: Peterson Institute for International Economics.

European Commission (2012), Decision of the European Parliament and of the Council Derogating Temporarily from Directive 2003/87/EC of the European Parliament and of the Council Establishing a Scheme for Greenhouse Gas Emission Allowance Trading within the Community, COM(2012) 697.

Fassbender, Bardo (1998), 'The United Nations Charter as Constitution of the International Community', *Columbia Journal of Transnational Law*, 36(3), 529–619.

Fauchald, Ole Kristian (2010), *International Environmental Governance: A Legal Analysis of Selected Options*, FNI Report 16/2010, Lysaker: Fridtjof Nansen Institute.

Fischer, Carolyn and Alan K. Fox (2009), *Comparing Policies to Combat Emissions Leakage: Border Tax Adjustments versus Rebates*, RFF Discussion Paper 09-02, Washington, DC: Resources for the Future.

Fischer, Carolyn and Richard G. Newell (2008), 'Environmental and Technology Policies for Climate Mitigation', *Journal of Environmental Economics and Management*, 55(2), 142–162.

Fischer-Lescano, Andreas and Gunther Teubner (2004), 'Regime-Collisions: the Vain Search for Legal Unity in the Fragmentation of Global Law', *Michigan Journal of International Law*, 25(4), 999–1046.

Fisher, Brian S., Melanie Ford, Guy Jakeman, Andrew Gurney, Jammie Penm, Anna Matysek and Don Gunasekera (2006), *Technological*

Development and Economic Growth, ABARE Research Report 06.1, Canberra: Australian Bureau of Agricultural and Resource Economics.

Fisher, Elizabeth, Bettina Lange, Eloise Scotford and Cinnamon Carlarne (2009), 'Maturity and Methodology: Starting a Debate about Environmental Law Scholarship', *Journal of Environmental Law*, 21(2), 213–250.

Fitzmaurice, Malgosia and Olefumi A. Elias (2005), *Contemporary Issues in the Law of Treaties*, Utrecht: Eleven.

Fujiwara, Noriko (2012), *Sector-specific Activities as the Driving Force towards a Low-Carbon Economy: From the Asia-Pacific Partnership to a Global Partnership*, Brussels: Centre for European Policy Studies.

Galiana, Isabel and Christopher Green (2009), 'Let the Global Technology Race Begin', *Nature*, 462(7273), 570–571.

Gardiner, Richard (2008), *Treaty Interpretation*, Oxford: Oxford University Press.

Gardner, Toby A., Neil D. Burgess, Naikoa Aguilar-Amuchastegui, Jos Barlow, Erika Berenguer, Tom Clements, Finn Danielsen, Joice Ferreira, Wendy Foden, Valerie Kapos, Saiful M. Khan, Alexander C. Lees, Luke Parry, Rosa Maria Roman-Cuesta, Christine B. Schmitt, Niels Strange, Ida Theilade and Ima C.G. Vieira (2012), 'A Framework for Integrating Biodiversity Concerns into National REDD+ Programmes', *Biological Conservation*, 154, 61–71.

Gehring, Thomas (2011), 'The Institutional Complex of Trade and Environment: Toward an Interlocking Governance Structure and a Division of Labor', in Sebastian Oberthür and Olav Schram Stokke (eds.), *Managing Institutional Complexity: Regime Interplay and Global Environmental Change*, Cambridge, MA: The MIT Press, 227–254.

Gehring, Thomas and Sebastian Oberthür (2006a), 'Comparative Empirical Analysis and Ideal Types of Institutional Interaction', in Sebastian Oberthür and Thomas Gehring (eds.), *Institutional Interaction in Global Environmental Governance. Synergy and Conflict among International and EU Policies*, Cambridge, MA: The MIT Press, 307–371.

Gehring, Thomas and Sebastian Oberthür (2006b), 'Introduction', in Sebastian Oberthür and Thomas Gehring (eds.), *Institutional Interaction in Global Environmental Governance. Synergy and Conflict among International and EU Policies*, Cambridge, MA: The MIT Press, 1–18.

Gehring, Thomas and Sebastian Oberthür (2009), 'The Causal Mechanisms of Interaction between International Institutions', *European Journal of International Relations*, 15(1), 125–156.

Gerstetter, Christiane, Dominic Marcellino and Elena von Sperber (2010), 'Technology Transfer in the International Climate Negotiations

– The State of Play and Suggestions for the Way Forward', *Carbon and Climate Law Review*, 3(1), 3–12.

Ghouri, Ahmad Ali (2012), 'Is Characterization of Treaties a Solution to Treaty Conflicts?', *Chinese Journal of International Law*, 11(2), 247–280.

Global Environment Facility (2010), *Implementation of the Poznan Strategic Program on Technology Transfer: Report of the GEF to the Sixteenth Session of the Conference of the Parties to the United Nations Framework Convention on Climate Change*, Washington, DC: Global Environment Facility.

Global Methane Initiative (2010), 'Terms of Reference of the Global Methane Initiative', available at http://www.globalmethane.org/documents/about_termsofreference_en.pdf (accessed 11 December 2013).

González-Calatayud, Alexandra and Gabrielle Marceau (2002), 'The Relationship between the Dispute-Settlement Mechanisms of MEAs and those of the WTO', *Review of European Community and International Environmental Law*, 11(3), 275–286.

Graichen, Patrick (2005), 'Can Forestry Gain from Emissions Trading? Rules Governing Sinks Projects Under the UNFCCC and the EU Emissions Trading System', *Review of European Community and International Environmental Law*, 14(1), 11–18.

Green, Andrew (2005), 'Climate Change, Regulatory Policy and the WTO: How Constraining Are Trade Rules?', *Journal of International Economic Law*, 8(1), 143–189.

Green, Andrew (2006), 'Trade Rules and Climate Change Subsidies', *World Trade Review*, 5(3), 377–414.

Grubb, Michael (2004), 'Technology Innovation and Climate Change Policy: An Overview of Issues and Options', *Keio Economic Studies*, 41(2), 103–132.

Grubb, Michael (2011), 'International Climate Finance from Border Carbon Cost Levelling', *Climate Policy*, 11(3), 1050–1057.

Grubb, Michael and Farhana Yamin (2001), 'Climatic Collapse at The Hague: What Happened, Why, and Where Do We Go from Here?' *International Affairs*, 77(2), 261–276.

Grubb, Michael, Christiaan Vrolijk and Duncan Brack (1999), *The Kyoto Protocol: A Guide and Assessment*, London, UK: Earthscan.

Gulbrandsen, Lars H. and Steinar Andresen (2004), 'NGO Influence in the Implementation of the Kyoto Protocol: Compliance, Flexibility Mechanisms, and Sinks', *Global Environmental Politics*, 4(4), 54–75.

Gupta, Joyeeta (1997), *The Climate Change Convention and Developing Countries – From Conflict to Consensus?*, PhD thesis, Vrije Universiteit Amsterdam.

Gupta, Joyeeta (1998), 'Leadership in the Climate Regime: Inspiring the Commitment of Developing Countries in the Post-Kyoto Phase', *Review of European Community and International Environmental Law*, 7(2), 178–188.

Gupta, Joyeeta and Harro van Asselt (2006), 'Helping Operationalise Article 2: A Transdisciplinary Methodological Tool for Evaluating When Climate Change Is Dangerous', *Global Environmental Change*, 16(1), 83–94.

Haas, Ernst B. (1980), 'Why Collaborate? Issue-Linkage and International Regimes', *World Politics*, 32(3), 357–402.

Haas, Peter M. (2004), 'Addressing the Global Governance Deficit', *Global Environmental Politics*, 4(4), 1–15.

Hafner, Gerhard (2000), 'Risks Ensuing from Fragmentation of International Law', in *Official Records of the General Assembly, Fifty-fifth session, Supplement No. 10*, UN Doc. A/55/10, 2000, Annex, 326–354.

Hafner, Gerhard (2004), 'Pros and Cons Ensuing from Fragmentation of International Law', *Michigan Journal of International Law*, 25(4), 849–863.

Hafner-Burton, Emilie, David G. Victor and Yonatan Lupu (2012), 'Political Science Research on International Law: The State of the Field', *American Journal of International Law*, 106(1), 47–97.

Haites, Erik, Duan Maosheng and Stephhen Seres (2006), 'Technology Transfer by CDM Projects', *Climate Policy*, 6(3), 327–344.

Hale, Thomas (2011), 'A Climate Coalition of the Willing', *The Washington Quarterly*, 34(1), 89–101.

Hale, Thomas (2013), *How the UNFCCC Can Drive Climate Ambition in Advance of a Treaty: Record, Review, Reinforce, Recruit*, Oxford: Blavatnik School of Government.

Halsnæs, Kirsten, Priyadarshi Shukla, Dilip Ahuja, Grace Akumu, Roger Beale, Jae Edmonds, Christian Gollier, Arnulf Grübler, Minh Ha Duong, Anil Markandya, Mack McFarland, Elena Nikitina, Taishi Sugiyama, Arturo Villavicencio and Ji Zou (2007), 'Framing Issues', in Bert Metz, Ogunlade Davidson, Peter Bosch, Rutu Dave and Leo Meyer (eds.), *Climate Change 2007: Mitigation. Contribution of Working Group III to the Fourth Assessment Report of the Intergovernmental Panel on Climate Change*, Cambridge, UK: Cambridge University Press, 117–167.

Hamilton, Katherine, Ricardo Bayon and Amanda Hawn (2008), 'Carving a Niche for Forests in the Voluntary Carbon Markets', in Charlotte Streck, Robert O'Sullivan, Toby Janson-Smith and Richard Tarasofsky (eds.), *Climate Change and Forests: Emerging Policy and Market Opportunities*, Washington, DC: Brookings Institution Press, 292–307.

Hardin, Garrett (1968), 'The Tragedy of the Commons', *Science*, 162(3859), 1243–1248.

Harrop, Stuart R. (2011), '"Living In Harmony With Nature"? Outcomes of the 2010 Nagoya Conference of the Convention on Biological Diversity', *Journal of Environmental Law*, 23(1), 117–128.

Harrop, Stuart R. and Diana J. Pritchard (2011), 'A Hard Instrument Goes Soft: The Implications of the Convention on Biological Diversity's Current Trajectory', *Global Environmental Change*, 21(2), 474–480.

Hart, H.L.A. (1994), *The Concept of Law*, 2nd edition, Oxford: Oxford University Press.

Harvey, Celia, Barney Dickson and Cyril Kormos (2010), Opportunities for Achieving Biodiversity Conservation through REDD, *Conservation Letters*, 3(1), 53–61.

Hasenclever, Andreas, Peter Mayer and Volker Rittberger (1997), *Theories of International Regimes*, Cambridge, UK: Cambridge University Press.

Haverkamp, Jennifer (2008), *International Aspects of a Climate Change Cap and Trade Program*, Testimony before the Committee on Finance, US Senate (14 February 2008), available at http://www.edf.org/sites/default/files/7644_Testimony_Finance_Senate.pdf (accessed 11 December 2013).

Helfer, Laurence R. (2004), 'Regime Shifting: The TRIPs Agreement and New Dynamics of International Intellectual Property Lawmaking', *Yale Journal of International Law*, 29(1), 1–83.

Helfer, Laurence R. (2009), 'Regime Shifting in the International Intellectual Property System', *Perspectives on Politics*, 7(1), 39–44.

Helm, Carsten and Detlef Sprinz (2000), 'Measuring the Effectiveness of International Environmental Regimes', *Journal of Conflict Resolution*, 44(5), 635–652.

Hertel, Michael (2011), 'Climate-Change-Related Trade Measures and Article XX: Defining Discrimination in Light of the Principle of Common but Differentiated Responsibilities', *Journal of World Trade*, 45(3), 653–678.

Hicks, Bethany Lukitsch (1999), 'Treaty Congestion in International Environmental Law: The Need for Greater International Coordination, Comment', *University of Richmond Law Review*, 32(4), 1643–1674.

Hillgenberg, Hartmut (1999), 'A Fresh Look at Soft Law', *European Journal of International Law*, 10(3), 499–515.

Hoffert, Martin I., Ken Caldeira, Gregory Benford, David R. Criswell, Christopher Green, Howard Herzog, Atul K. Jain. Haroon S. Kheshgi, Klaus S. Lackner, John S. Lewis, H. Douglas Lightfoot, Wallace Manheimer, John C. Mankins, Michael E. Mauel, L. John Perkins,

Michael E. Schlesinger, Tyler Volk and Tom M.L. Wigley (2002), 'Advanced Technology Paths to Global Climate Stability: Energy for a Greenhouse Planet', *Science*, 298(5595), 981–987.

Hoffman, Matthew J. (2011), *Climate Governance at the Crossroads: Experimenting with a Global Response after Kyoto*, Oxford: Oxford University Press.

Holmes, Peter, Tom Reilly and Jim Rollo (2011), 'Border Carbon Adjustments and the Potential for Protectionism', *Climate Policy*, 11(2), 883–900.

Honkonen, Tuula (2009), *The Common but Differentiated Responsibility Principle in Multilateral Environmental Agreements: Regulatory and Policy Aspects*, Leiden: Kluwer Law International.

Hooghe, Liesbet and Gary Marks (2003), 'Unraveling the Central State, but How? Types of Multi-level Governance', *American Political Science Review*, 97(2), 233–243.

Hourcade, Jean-Charles, Damien Demailly, Karsten Neuhoff and Misato Sato (2008), *Differentiation and Dynamics of EU ETS Industrial Competitiveness*, Cambridge, UK: Climate Strategies.

Houser, Trevor, Rob Bradley, Britt Childs, Jacob Werksman and Robert Heilmayr (2008), *Leveling the Carbon Playing Field: International Competition and U.S. Climate Policy Design*, Washington, DC: Peterson Institute for International Economics/World Resources Institute.

Hovi, Jon, Detlef F. Sprinz and Arild Underdal (2003), 'The Oslo-Potsdam Solution to Measuring Regime Effectiveness: Critique, Response, and the Road Ahead', *Global Environmental Politics*, 3(3), 74–96.

Howse, Robert (2007), *The Use and Abuse of "Other Relevant Rules of International Law" in Treaty Interpretation: Insights from WTO Trade/Environment Litigation*, Institute for International Law and Justice Working Paper 2007/1, New York, NY: New York University Institute for International Law and Justice.

Howse, Robert and Antonia L. Eliason (2009), 'Domestic and International Strategies to Address Climate Change: An Overview of the WTO Legal Issues', in Thomas Cottier, Olga Nartova and Sadeq Z. Bigdeli (eds.), *International Trade Regulation and the Mitigation of Climate Change*, Cambridge, UK: Cambridge University Press, 48–93.

Howse, Robert and Donald Regan (2000), 'The Product/Process Distinction. An Illusionary Basis for Disciplining "Unilateralism" in Trade Policy', *European Journal of International Law*, 11(2), 249–289.

Howse, Robert and Ruti Teitel (2010), 'Beyond Compliance: Rethinking Why International Law Really Matters', *Global Policy*, 1(2), 127–136.

Hufbauer, Gary Clyde, Steve Charnovitz and Jisun Kim (2009), *Global Warming and the World Trading System*, Washington, DC: Peterson Institute for International Economics.
Hulme, Mike (2009), *Why We Disagree About Climate Change: Understanding Controversy, Inaction and Opportunity*, Cambridge, UK: Cambridge University Press.
Humphreys, David (1996), *Forest Politics: The Evolution of International Cooperation*, London, UK: Earthscan.
Humphreys, David (2006), *Logjam: Deforestation and the Crisis of Global Governance*, London, UK: Earthscan.
Humphreys, Stephen (2012), 'Structural Ambiguity: Technology Transfer in Three Regimes', in Margaret A. Young (ed.), *Regime Interaction in International Law: Facing Fragmentation*, Cambridge, UK: Cambridge University Press, 175–198.
Humphreys, Stephen (ed.) (2009), *Human Rights and Climate Change*, Cambridge, UK: Cambridge University Press.
Hurrell, Andrew (2000), 'Conclusion: International Law and the Changing Constitution of International Society', in Michael Byers (ed.), *The Role of Law in International Politics*, Oxford: Oxford University Press, 327–347.
ILC (2000), *Official Records of the General Assembly, Fifty-fifth Session, Supplement No. 10*, UN Doc. A/55/10.
ILC (2006), *Fragmentation of International Law: Difficulties Arising from the Diversification and Expansion of International Law. Report of the Study Group of the International Law Commission*, UN Doc. A/CN.4/L.682 (13 April 2006).
Inomata, Tadanori (2008), *Management Review of Environmental Governance in the United Nations System*, UN Doc. JIU/REP/2008/3, Geneva: Joint Inspection Unit.
International Centre for Trade and Sustainable Development (2007), 'Disagreement over Biofuels, Climate-friendly Products', available at http://ictsd.org/i/news/bridges/3153/ (accessed 11 December 2013).
International Centre for Trade and Sustainable Development (2008), *Climate Change, Technology Transfer and Intellectual Property Rights*, Geneva: International Centre for Trade and Sustainable Development.
International Centre for Trade and Sustainable Development (2011), 'India Proposes Discussion of Trade Issues for Durban Climate Meet', *Bridges Weekly*, 15(32).
International Energy Agency (2003), *IEA Implementing Agreements. Background and Framework as of 2003*, Paris: International Energy Agency.

IPCC (2000), *Methodological and Technological Issues in Technology Transfer. Summary for Policymakers*, IPCC Special Report. Geneva: IPCC.

IPCC (2002), *Climate Change and Biodiversity*, IPCC Technical Paper V, Geneva: IPCC.

IPCC (2007), *Climate Change 2007: Synthesis Report. A Contribution of Working Groups I, II, and III to the Fourth Assessment Report of the Intergovernmental Panel on Climate Change*, Cambridge, UK: Cambridge University Press.

IPCC (2013), *Climate Change 2013: The Physical Science Basis. Working Group I Contribution to the Fifth Assessment Report of the Intergovernmental Panel on Climate Change. Summary for Policymakers*, Cambridge, UK: Cambridge University Press.

Ismer, Roland and Karsten Neuhoff (2007), 'Border Tax Adjustment: A Feasible Way to Support Stringent Emission Trading', *European Journal of Law and Economics*, 24(2), 137–164.

Ivanova, Maria (2012), 'Institutional Design and UNEP Reform: Historical Insights on Form, Function and Financing', *International Affairs*, 88(3), 565–584.

Jacquemont, Frédéric and Alejandro Caparrós (2002), 'The Convention on Biological Diversity and the Climate Change Convention 10 Years after Rio: Towards a Synergy of the Two Regimes?', *Review of European Community and International Environmental Law*, 11(2), 139–180.

Jaffe, Adam, Richard Newell and Robert N. Stavins (2005), 'A Tale of Two Market Failures: Technology and Environmental Policy', *Ecological Economics*, 54(2–3), 164–174.

Jagers, Sverker and Johannes Stripple (2003), 'Climate Governance beyond the State', *Global Governance*, 9(3), 385–399.

Jenks, Wilfred (1953), 'The Conflict of Law-Making Treaties', *British Yearbook of International Law*, 30, 401–453.

Jinnah, Sikina (2003), 'Emissions Trading under the Kyoto Protocol: NAFTA and WTO Concerns', *Georgetown International Environmental Law Review*, 15(4), 709–761.

Jinnah, Sikina (2010), 'Overlap Management in the World Trade Organization: Secretariat Influence on Trade-Environment Politics', *Global Environmental Politics*, 10(2), 64–79.

Jinnah, Sikina (2011), 'Marketing Linkages: Secretariat Governance of the Climate–Biodiversity Interface', *Global Environmental Politics*, 11(3), 23–43.

Jinnah, Sikina and Miquel Muñoz Cabré (eds.) (2011), 'Special Issue: Climate Bandwagoning: The Impacts of Strategic Linkages on Regime Design, Maintenance, and Death', *Global Environmental Politics*, 11(3).

Joint Liaison Group (2009), Report of the Meeting of the Joint Liaison Group of the Convention on Biological Diversity, the United Nations Convention to Combat Desertification, and the United Nations Framework Convention on Climate Change (New York, 14 May 2009).

Joint Liaison Group (2011), Eleventh Meeting of the Joint Liaison Group of the Rio Conventions (Bonn, 11 April 2011).

Jordan, Andrew and Adriaan Schout (2006), *The Coordination of the European Union: Exploring the Capacities of Networked Governance*, Oxford: Oxford University Press.

Jordan, Andrew, Rüdiger K.W. Wurzel and Anthony R. Zito (2005), 'The Rise of "New" Policy Instruments in Comparative Perspective: Has Governance Eclipsed Government?', *Political Studies*, 53(3), 477–496.

Jupille, Joseph and Duncan Snidal (2006), 'The Choice of International Institutions: Cooperation, Alternatives and Strategies', available at http://papers.ssrn.com/sol3/papers.cfm?abstract_id=1008945 (accessed 11 December 2013).

Justus, Debra and Cédric Philibert (2005), *International Energy Technology Collaboration and Climate Change Mitigation. Synthesis Report*, Paris: International Energy Agency.

Karlsson-Vinkhuyzen, Sylvia and Jeffrey S. McGee (2013), 'Legitimacy in an Era of Fragmentation: The Case of Global Climate Governance', *Global Environmental Politics*, 13(3), 56–78.

Karlsson-Vinkhuyzen, Sylvia and Harro van Asselt (2009), 'Introduction: Exploring and Explaining the Asia-Pacific Partnership on Clean Development and Climate', *International Environmental Agreements: Politics, Law and Economics*, 9(3), 195–211.

Karousakis, Katia (2009), *Promoting Biodiversity Co-Benefits in REDD*, OECD Environment Working Papers No. 11, Paris: OECD.

Keohane, Robert O. (1989), 'Neoliberal Institutionalism: A Perspective on World Politics', in Robert O. Keohane (ed.), *International Institutions and State Power. Essays in International Relations Theory*, Boulder, CO: Westview Press, 1–20.

Keohane, Robert O. (1997), 'International Relations and International Law: Two Optics', *Harvard International Law Journal*, 38(2), 487–502.

Keohane, Robert O. and David G. Victor (2011), 'The Regime Complex for Climate Change', *Perspectives on Politics*, 9(1), 7–23.

Khalastchi, Ruth and Ruth Mackenzie (1999), 'The Conservation and Sustainable Use of Forest Biological Diversity: The Role of the

Convention on Biological Diversity', in Richard G. Tarasofsky (ed.), *Assessing the International Forest Regime*, Gland: IUCN, 39–64.

Khor, Martin (2009), 'Moves to Tax South's Imports on Climate Grounds are Unfair', Third World Network Briefing Paper 1, UN Climate Change Talks, Intersessional Informal Consultations, Bonn, 10 August–14 August 2009.

Khrebtukova, Alexandra (2008), 'A Call to Freedom: Towards a Philosophy of International Law in an Era of Fragmentation', *Journal of International Law and International Relations*, 4(1), 51–103.

Kingsbury, Benedict (1999), 'Is the Proliferation of International Courts and Tribunals a Systemic Problem?', *New York University Journal of International Law and Politics*, 31(4), 679–696.

Kingsbury, Benedict, Nico Krisch and Richard Stewart (2005), 'The Emergence of Global Administrative Law', *Law and Contemporary Problems*, 68(3–4), 15–61.

Klabbers, Jan (1996), 'The Redundancy of Soft Law', *Nordic Journal of International Law*, 65(2), 167–182.

Klabbers, Jan (2002), *An Introduction to International Institutional Law*, Cambridge, UK: Cambridge University Press.

Klabbers, Jan (2005), 'The Relative Autonomy of International Law or the Forgotten Politics of Interdisciplinarity', *Journal of International Law and International Relations*, 1(1–2), 35–48.

Klabbers, Jan (2008), *Treaty Conflict and the European Union*, Cambridge, UK: Cambridge University Press.

Klabbers, Jan, Anne Peters and Geir Ulfstein (2009), *The Constitutionalization of International Law*, Oxford: Oxford University Press.

Kornicker Uhlmann, Eva M. (1998), 'State Community Interests, Jus Cogens and Protection of the Global Environment: Developing Criteria for Peremptory Norms', *Georgetown International Environmental Law Review*, 11(1), 101–135.

Koskenniemi, Martti (1992), 'Breach of Treaty or Non-Compliance: Reflections on the Enforcement of the Montreal Protocol', *Yearbook of International Environmental Law*, 3, 123–162.

Koskenniemi, Martti (2000), 'Carl Schmitt, Hans Morgenthau, and the Image of Law in International Relations', in Michael Byers (ed.), *The Role of Law in International Politics: Essays in International Relations and International Law*, Oxford: Oxford University Press, 17–34.

Koskenniemi, Martti (2004), 'Global Governance and Public International Law', *Kritische Justiz*, 37, 241–254.

Koskenniemi, Martti (2005), *From Apology to Utopia: The Structure of International Legal Argument*, Cambridge, UK: Cambridge University Press.

Koskenniemi, Martti (2007), 'The Fate of Public International Law: Between Technique and Politics', *The Modern Law Review*, 70(1), 1–30.

Koskenniemi, Martti (2009), 'Miserable Comforters: International Relations as New Natural Law', *European Journal of International Relations*, 15(3), 395–422.

Koskenniemi, Martti (2012), 'Hegemonic Regimes', in Margaret A. Young (ed.), *Regime Interaction in International Law: Facing Fragmentation*, Cambridge, UK: Cambridge University Press, 305–324.

Koskenniemi, Martti and Päivi Leino (2002), 'Fragmentation of International Law? Postmodern Anxieties', *Leiden Journal of International Law*, 15(3), 553–579.

Kossoy, Alexandre and Philippe Ambrosi (2010), *State and Trends of the Carbon Market 2010*, Washington, DC: World Bank.

Krasner, Stephen D. (1983), 'Structural Causes and Regime Consequences: Regimes as Intervening Variables', in Stephen Krasner (ed.), *International Regimes*, Ithaca, NY: Cornell University Press, 1–21.

Krisch, Nico (2010), *Beyond Constitutionalism: The Pluralist Structure of Postnational Law*, Oxford: Oxford University Press.

Kulovesi, Kati (2011a), '"Make Your Own Special Song, even if Nobody Else Sings Along": International Aviation Emissions and the EU Emissions Trading Scheme', *Climate Law*, 2(4), 535–558.

Kulovesi, Kati (2011b), *The WTO Dispute Settlement System. Challenges of the Environment, Legitimacy and Fragmentation*, Dordrecht: Kluwer Law International.

Lamy, Pascal (2007), 'Doha Could Deliver Double-Win for Environment and Trade. Speech at the Informal Trade Ministers' Dialogue on Climate Change in Bali on 8–9 December 2007', available at http://www.wto.org/english/news_e/sppl_e/sppl83_e.htm (accessed 11 December 2013).

Lang, Andrew T.F. (2011), *World Trade Law after Neoliberalism: Re-imagining the Global Economic Order*, Oxford: Oxford University Press.

Lang, Andrew T.F. (2012), 'Legal Regimes and Professional Knowledges: The Internal Politics of Regime Definition', in Margaret A. Young (ed.), *Regime Interaction in International Law: Facing Fragmentation*, Cambridge, UK: Cambridge University Press, 113–135.

Lawlor, Kathleen, Aaron Jenkins, Lydia P. Olander and Brian C. Murray (2010), *Expanding the Scope of International Terrestrial Carbon Options: Implications of REDD+ and Beyond*, Durham, NC: Nicholas Institute for Environmental Policy Solutions.

Lawrence, Peter (2007), 'The Asia Pacific Partnership on Clean Development and Climate (AP6): A Distraction to the Kyoto Process or a

Viable Alternative?', *Asia Pacific Journal of Environmental Law*, 10(3/4), 183–209.

Lawrence, Peter (2009), 'Australian Climate Policy and the Asia Pacific Partnership on Clean Development and Climate (APP): From Howard to Rudd – Continuity or Change?', *International Environmental Agreements: Politics, Law and Economics*, 9(3), 281–299.

Lazarus, Richard J. (2009), 'Super Wicked Problems and Climate Change: Restraining the Present to Liberate the Future', *Cornell Law Review*, 94(5), 1153–1233.

Le Prestre, Philippe G. (2002), 'The CBD at Ten: The Long Road to Effectiveness', *Journal of International Wildlife Law and Policy*, 5(3), 269–285.

Leebron, David W. (2002), 'Linkages', *American Journal of International Law*, 96(1), 5–27.

Levin, Kelly, Steven Bernstein, Benjamin Cashore and Graeme Auld (2007), 'Playing it Forward: Path Dependency, Progressive Incrementalism, and the "Super Wicked" Problem of Global Climate Change', paper prepared for the International Studies Association Annual Convention, Chicago, IL, 28 February–3 March 2007.

Levin, Kelly, Constance McDermott and Benjamin Cashore (2008), 'The Climate Regime as Global Forest Governance: Can Reduced Emissions from Deforestation and Forest Degradation (REDD) Initiatives Pass a "Dual Effectiveness" Test?', *International Forestry Review*, 10(3), 538–549.

Levin, Kelly, Benjamin Cashore and Jonathan Koppell (2009), 'Can Non-State Certification Systems Bolster State-centered Efforts to Promote Sustainable Development through the Clean Development Mechanism?', *Wake Forest Law Review*, 44(3), 777–798.

Levy, Marc A., Oran R. Young and Michael Zürn (1995), 'The Study of International Regimes', *European Journal of International Relations*, 1(3), 267–330.

Linderfalk, Ulf (2008), 'Who Are "the Parties"? Article 31, Paragraph 3(c) of the Vienna Convention and the "Principle of Systemic Integration" Revisited', *Netherlands International Law Review*, 55(3), 343–364.

Lindroos, Anja (2005), 'Addressing Norm Conflicts in a Fragmented Legal System', *Nordic Journal of International Law*, 74(1), 27–66.

Lindroos, Anja and Michael Mehling (2005), 'Dispelling the Chimera of "Self-Contained Regimes". International Law and the WTO', *European Journal of International Law*, 16(5), 857–877.

Lindroos, Anja and Michael Mehling (2008), 'From Autonomy to Integration? International Law, Free Trade and the Environment', *Nordic Journal of International Law*, 77(3), 253–273.

Littleton, Matthew (2008), *The TRIPS Agreement and Transfer of Climate-Change-Related Technologies to Developing Countries*, DESA Working Paper No. 71. Doc. No. ST/ESA/2008/DWP/71, New York, NY: United Nations Department of Economic and Social Affairs.

Long, Andrew (2011), 'Global Climate Governance to Enhance Biodiversity and Well-Being: Integrating Non-State Networks and Public International Law in Tropical Forests', *Environmental Law*, 41(1), 95–164.

Lovell, Heather (2010), 'Governing the Carbon Offset Market', *WIREs Climate Change*, 1(3), 353–362.

Low, Bobbi, Elinor Ostrom, Carl Simon and James Wilson (2003), 'Redundancy and Diversity: Do They Influence Optimal Management?', in Fikret Berkes, Johan Colding and Carl Folke (eds.), *Navigating Social-Ecological Systems: Building Resilience for Complexity and Change*, Cambridge, UK: Cambridge University Press, 83–114.

Lowe, Vaughan (1999), 'Sustainable Development and Unsustainable Arguments', in Alan Boyle and David Freestone (eds.), *International Law and Sustainable Development: Past Achievements and Future Challenges*, Oxford: Oxford University Press, 19–38.

Mann, Michael E. (2009), 'Defining Dangerous Anthropogenic Interference', *Proceedings of the National Academy of Sciences*, 106(11), 65–66.

Marceau, Gabrielle (1999), 'A Call for Coherence in International Law – Praises for the Prohibition against "Clinical Isolation" in WTO Dispute Settlement', *Journal of World Trade*, 33(5), 87–152.

Marceau, Gabrielle (2001), 'Conflicts of Norms and Conflicts of Jurisdictions: The Relationship between the WTO Agreement and MEAs and Other Treaties', *Journal of World Trade*, 35(6), 1081–1131.

Marschik, Axel (1998), 'Too Much Order? The Impact of Special Secondary Norms on the Unity and Efficacy of the International Legal System', *European Journal of International Law*, 9(1), 212–239.

Martin, Marisa A. (2007), 'Trade Law Implications of Restricting Participation in the European Union Emissions Trading Scheme', *Georgetown International Environmental Law Review*, 19(3), 437–474.

Martineau, Anne-Charlotte (2009), 'The Rhetoric of Fragmentation: Fear and Faith in International Law', *Leiden Journal of International Law*, 22(1), 1–28.

Matsumura, Hiroshi (2000), *Japan and the Kyoto Protocol. Conditions for Ratification*, London, UK: The Royal Institute of International Affairs.

Matz-Lück, Nele (2008a), 'Conflicts between Treaties', in Rüdiger Wolfrum (ed.), *The Max Planck Encyclopedia of Public International Law*, Oxford: Oxford University Press.

Matz-Lück, Nele (2008b), 'Harmonization, Systemic Integration and "Mutual Supportiveness" as Conflict-Solution Techniques – Different Modes of Interpretation as a Challenge to Negative Effects of Fragmentation?', *Finnish Yearbook of International Law*, 17, 39–53.

Matz-Lück, Nele (2012), 'Norm Interpretation across International Regimes: Competences and Legitimacy', in Margaret A. Young (ed.), *Regime Interaction in International Law: Facing Fragmentation*, Cambridge, UK: Cambridge University Press, 201–234.

McAdam, Jane (2012), *Climate Change, Forced Migration, and International Law*, Oxford: Oxford University Press.

McCabe, Daniel G. (2007), 'Resolving Conflicts between Multilateral Environmental Agreements: The Case of the Montreal and Kyoto Protocols', *Fordham Environmental Law Review*, 18(2), 433–466.

McDermott, Constance, Aran O'Caroll and Peter Wood (2007), *International Forest Policy – The Instruments, Agreements and Processes that Shape It*, New York, NY: UN Department of Economic and Social Affairs.

McGee, Jeffrey and Ros Taplin (2006), 'The Asia-Pacific Partnership on Clean Development and Climate: A Complement or Competitor to the Kyoto Protocol?', *Global Change, Peace & Security*, 18(3), 173–192.

McGee, Jeffrey and Ros Taplin (2008), 'The Asia-Pacific Partnership and the United States' International Climate Change Policy', *Colorado Journal of International Environmental Law and Policy*, 19(2), 179–218.

McGee, Jeffrey and Ros Taplin (2009a), 'The Asia-Pacific Partnership on Clean Development and Climate: A Retreat from the Principle of Common but Differentiated Responsibilities?', *McGill Journal of Sustainable Development Law and Policy*, 5(1), 11–43.

McGee, Jeffrey and Ros Taplin (2009b), 'The Role of the Asia-Pacific Partnership in Discursive Contestation of the International Climate Regime', *International Environmental Agreements: Politics, Law and Economics*, 9(3), 213–238.

McGee, Jeffrey and Ros Taplin (2012), 'The Asia-Pacific Partnership: A Deepened Market Liberal Model for the International Climate Regime?', in Brad Jessup and Kim Rubinstein (eds.), *Environmental Discourses in Public and International Law*, Cambridge, UK: Cambridge University Press, 308–330.

McKenzie, Michael (2008), 'Climate Change and the Generalized System of Preferences', *Journal of International Economic Law*, 11(3), 679–695.

McLachlan, Campbell (2005), 'The Principle of Systemic Integration and Article 31(3)(c) of the Vienna Convention', *International and Comparative Law Quarterly*, 54(2), 279–319.

McNeely, Jeffrey A. (2003), 'Energy and Biodiversity: Understanding Complex Relationships', in Adrian J. Bradbrook and Richard L. Ottinger (eds.), *Energy Law and Sustainable Development*, Gland: IUCN, 31–44.

Mee, Laurence D., Holly T. Dublin and Anton A. Eberhard (2008), 'Evaluating the Global Environment Facility: A Goodwill Gesture or a Serious Attempt to Deliver Global Benefits?', *Global Environmental Change*, 18(4), 800–810.

Mehling, Michael A. (2004), 'Betwixt Scylla and Charybdis: the Concept of Effectiveness in International Environmental Law', *Finnish Yearbook of International Law*, 13, 129–182.

Meinhausen, Malte (2006), 'What Does a 2°C Target Mean for Greenhouse Gas Concentrations? A Brief Analysis Based on Multi-Gas Emission Pathways and Several Climate Sensitivity Uncertainty Estimates', in Hans-Joachim Schellnhuber, Wolfgang Cramer, Nebojsa Nakicenovic, Tom Wigley and Gary Yohe (eds.), *Avoiding Dangerous Climate Change*, Cambridge, UK: Cambridge University Press, 265–280.

Meinshausen, Malte and Bill Hare (2003), 'Sinks in the CDM: After the Climate, Biodiversity Goes Down the Drain. An Analysis of the CDM Sinks Agreement at CoP-9', available at http://www.greenpeace.org/international/Global/international/planet-2/report/2006/3/sinks-in-the-cdm-after-the-cl-2.pdf (accessed 11 December 2013).

Meinshausen, Malte, Nicolai Meinshausen, William Hare, Sarah C.B. Raper, Katja Frieler, Reto Knutti, David J. Frame and Myles R. Allen (2009), 'Greenhouse-Gas Emission Targets for Limiting Global Warming to 2°C', *Nature*, 458(7242), 1158–1162.

Merger, Eduard, Michael Dutschke and Louis Verchot (2011), 'Options for REDD+ Voluntary Certification to Ensure Net GHG Benefits, Poverty Alleviation, Sustainable Management of Forests and Biodiversity Conservation', *Forests*, 2, 550–577.

Michaels, Ralf and Joost Pauwelyn (2011), 'Conflict of Norms or Conflict of Laws? Different Techniques in the Fragmentation of International Law', in Tomer Broude and Yuval Shany (eds.), *Multi-Sourced Equivalent Norms in International Law*, Oxford: Hart, 19–44.

Miles, Edward L., Arild Underdal, Steinar Andresen, Jørgen Wettestad, Jon Birger Skjærseth and Elaine M. Carlin (eds.) (2002), *Environmental Regime Effectiveness: Confronting Theory with Evidence*, Cambridge, MA: The MIT Press.

Millennium Ecosystem Assessment (2005), *Ecosystems and Human Well-being: Synthesis*, Washington, DC: Island Press.

Mitchell, Ronald B. (2008), 'Evaluating the Performance of International Institutions: What to Evaluate and How to Evaluate It', in Oran R.

Young, Leslie A. King and Heike Schroeder (eds.), *Institutions and Environmental Change: Principal Findings, Applications, and Research Frontiers*, Cambridge, MA: The MIT Press, 79–114.

Moncel, Remi and Harro van Asselt (2012), 'All Hands on Deck! Mobilizing Climate Change Action beyond the UNFCCC', *Review of European Community and International Environmental Law*, 21(3), 163–176.

Morgera, Elisa (2011), 'Far Away, So Close: A Legal Analysis of the Increasing Interactions between the Convention on Biological Diversity and Climate Change Law', *Climate Law*, 2(1), 85–115.

Motaal, Doaa Abdel (2002), 'The Observership of Intergovernmental Organizations in the WTO, Post-Doha: Is there Political Will to Bridge the Divide?', *Journal of World Intellectual Property*, 5(3), 477–489.

Murphy, Deborah, John Van Ham and John Drexhage (2005), *Climate Change and Technology*, Winnipeg: International Institute for Sustainable Development.

Naím, Moisés (2009), 'Minilateralism. The Magic Number to Get Real International Action', *Foreign Policy*, July/August, 5–8.

Neuhoff, Karsten (2011), *Climate Policy after Copenhagen: The Role of Carbon Pricing*, Cambridge, UK: Cambridge University Press.

Neuhoff, Karsten, Roland Ismer and Vice Yu (2008), 'International Cooperation to Limit the Use of Border Adjustment, Workshop Summary', South Center, Geneva, September 10, 2008, Cambridge, UK: Climate Strategies.

New Zealand Ministry of Foreign Affairs (undated), Asia-Pacific Partnership on Clean Development and Climate, available at http://www.mfe.govt.nz/issues/climate/resources/cabinet-papers/mfat-cab-paper.html (accessed 11 December 2013).

O'Neill, Brian C. and Michael Oppenheimer (2002), 'Dangerous Climate Impacts and the Kyoto Protocol', *Science*, 296(5575), 1971–1972.

O'Sullivan, Robert (2008), 'Reducing Emissions from Deforestation in Developing Countries: An Introduction', in Charlotte Streck, Charlotte, Robert O'Sullivan, Toby Janson-Smith and Richard Tarasofsky (eds.), *Climate Change and Forests: Emerging Policy and Market Opportunities*, Washington, DC: Brookings Institution Press, 179–190.

Oberthür, Sebastian (2001), *The International Regime on Climate Change*, Project Deliverable No. D 2 of the Project 'Institutional Interaction – How to Prevent Conflicts and Enhance Synergies Between International and European Environmental Institutions', Berlin: Ecologic.

Oberthür, Sebastian (2002), 'Clustering of Multilateral Environmental Agreements: Potential and Limitations', *International Environmental Agreements: Politics, Law and Economics*, 2(4), 317–340.

Oberthür, Sebastian (2006), 'The Climate Change Regime: Interactions with ICAO, IMO, and the EU Burden-Sharing Agreement', in Sebastian Oberthür and Thomas Gehring (eds.), *Institutional Interaction in Global Environmental Governance. Synergy and Conflict among International and EU Policies*, Cambridge, MA: The MIT Press, 53–77.

Oberthür, Sebastian (2009), 'Interplay Management: Enhancing Environmental Policy Integration among International Institutions', *International Environmental Agreements: Politics, Law and Economics*, 9(4), 371–391.

Oberthür, Sebastian and Thomas Gehring (2004), 'Reforming International Environmental Governance: An Institutionalist Critique of the Proposal for a World Environment Organisation', *International Environmental Agreements: Politics, Law and Economics*, 4(4), 349–381.

Oberthür, Sebastian and Thomas Gehring (2006a), 'Conceptual Foundations of Institutional Interaction', in Sebastian Oberthür and Thomas Gehring (eds.), *Institutional Interaction in Global Environmental Governance: Synergy and Conflict among International and EU Policies*, Cambridge, MA: The MIT Press, 19–52.

Oberthür, Sebastian and Thomas Gehring (eds.) (2006b), *Institutional Interaction in Global Environmental Governance. Synergy and Conflict among International and EU Policies*, Cambridge, MA: The MIT Press.

Oberthür, Sebastian and Thomas Gehring (2011), 'Institutional Interaction: Ten Years of Scholarly Development', in Sebastian Oberthür and Olav Schram Stokke (eds.), *Managing Institutional Complexity: Regime Interplay and Global Environmental Change*, Cambridge, MA: The MIT Press, 25–58.

Oberthür, Sebastian and Hermann E. Ott (1999), *The Kyoto Protocol. International Climate Policy for the 21st Century*, Berlin: Springer.

Oberthür, Sebastian and Olav Schram Stokke (2011a), 'Conclusions: Decentralized Interplay Management in an Evolving Interinstitutional Order', in Sebastian Oberthür and Olav Schram Stokke (eds.), *Managing Institutional Complexity: Regime Interplay and Global Environmental Change*, Cambridge, MA: The MIT Press, 313–341.

Oberthür, Sebastian and Olav Schram Stokke (eds.) (2011b), *Managing Institutional Complexity: Regime Interplay and Global Environmental Change*, Cambridge, MA: The MIT Press.

Oberthür, Sebastian and Olav Schram Stokke (2012). 'Institutional Complexity and Interplay Management: Compatibility and Change in Global Governance', paper submitted to the ISA Annual Convention in San Diego, USA, April 2012, and to the Earth System Governance conference in Lund, Sweden, April 212.

Oberthür, Sebastian, Claire Dupont and Yasuko Matsumoto (2011), 'Managing Policy Contradictions Between the Montreal and Kyoto Protocols: The Case of Fluorinated Greenhouse Gases', in Sebastian Oberthür and Olav Schram Stokke (eds.), *Managing Institutional Complexity: Regime Interplay and Global Environmental Change*, Cambridge, MA: The MIT Press, 115–142.

Ockwell, David G., Ruediger Haum, Alexandra Mallett and Jim Watson (2010), 'Intellectual Property Rights and Low Carbon Technology Transfer: Conflicting Discourses of Diffusion and Development', *Global Environmental Change*, 20(4), 729–739.

Orakhelashvili, Alexander (2008), 'The Interaction between Human Rights and Humanitarian Law: Fragmentation, Conflict, Parallelism or Convergence?', *European Journal of International Law*, 19(1), 161–182.

Orellana, Marcos A. (2002), 'The Swordfish Dispute between the EU and Chile at the ITLOS and the WTO', *Nordic Journal of International Law*, 71(1), 55–81.

Orsini, Amandine (2013), 'Multi-Forum Non-State Actors: Navigating the Regime Complexes for Forestry and Genetic Resources', *Global Environmental Politics*, 13(3), 34–55.

Ostrom, Elinor (2010), 'Polycentric Systems for Coping with Collective Action and Global Environmental Change', *Global Environmental Change*, 20(4), 550–557.

Ostrom, Vincent, Charles M. Tiebout and Robert Warren (1961), 'The Organization of Government in Metropolitan Areas: A Theoretical Inquiry', *American Political Science Review*, 55(4), 831–842.

Pacala, Stephen and Rob Socolow (2004), 'Stabilization Wedges: Solving the Climate Problem for the Next 50 Years with Current Technologies', *Science*, 305(5686), 968–972.

Parmesan, Camilla and Gary Yohe (2003), 'A Globally Coherent Fingerprint of Climate Change Impacts across Natural Systems', *Nature*, 421(6918), 37–42.

Pattberg, Philipp and Johannes Stripple (2008), 'Beyond the Public and Private Divide: Remapping Transnational Climate Governance in the 21st Century', *International Environmental Agreements: Politics, Law and Economics*, 8(4), 367–388.

Pauwelyn, Joost (2002), 'The Use of Experts in WTO Dispute Settlement', *International and Comparative Law Quarterly*, 51(2), 325–364.

Pauwelyn, Joost (2003a), *Conflict of Norms in Public International Law. How WTO Law Relates to other Rules of International Law*, Cambridge, UK: Cambridge University Press.

Pauwelyn, Joost (2003b), 'How to Win a WTO Dispute Based on Non-WTO Law? Questions of Jurisdiction and Merits', *Journal of World Trade*, 37(6), 907–952.
Pauwelyn, Joost (2006), 'Fragmentation of International Law', in Rüdiger Wolfrum (ed.), *The Max Planck Encyclopedia of Public International Law*, Oxford: Oxford University Press.
Pauwelyn, Joost (2007), *US Federal Climate Policy and Competitiveness Concerns: The Limits and Options of International Trade Law*, Durham, NC: The Nicholas Institute for Environmental Policy Solutions.
Pauwelyn, Joost (2012), 'Is It International Law or Not and Does It Even Matter?', in Joost Pauwelyn, Ramses Wessel and Jan Wouters (eds.), *Informal International Lawmaking*, Oxford: Oxford University Press, 125–161.
Pauwelyn, Joost (2013), 'The End of Differential Treatment for Developing Countries? Lessons from the Trade and Climate Change Regimes', *Review of European, Comparative and International Environmental Law*, 22(1), 29–41.
Pavoni, Riccardo (2010), 'Mutual Supportiveness as a Principle of Interpretation and Law-Making: A Watershed for the "WTO-and-Competing-Regimes" Debate?', *European Journal of International Law*, 21(3), 649–667.
Pearce, Fred (2005), 'US-led Emissions Pact Seen as Kyoto Rival', *New Scientist* (28 July 2005), available at http://www.newscientist.com/article/dn7744 (accessed 11 December 2013).
Peay, Sarah (2007), 'Joining the Asia-Pacific Partnership: The Environmentally Sound Decision?', *Colorado Journal of International Environmental Law and Policy*, 18(2), 477–512.
Persson, Sofia (2010), *Practical Aspects of Border Carbon Adjustments: Using a Trade Facilitation Perspective to Assess Trade Cost*, Geneva: International Centre for Trade and Sustainable Development.
Petsonk, Annie (1999), 'The Kyoto Protocol and the WTO: Integrating Greenhouse Gas Emission Allowance Trading into the Global Marketplace', *Duke Environmental Law and Policy Forum*, 10(1), 185–220.
Pezzey, John C.V., Frank Jotzo and John Quiggin (2008), 'Fiddling while the Carbon Burns: Why Climate Policy Needs Pervasive Emission Pricing as well as Technology Promotion', *The Australian Journal of Agricultural and Resource Economics*, 52(1), 97–110.
Pistorius, Till, Christine B. Schmitt, Dinah Benick and Steffen Entenmann (2010), *Greening REDD+: Challenges and Opportunities for Forest Biodiversity Conservation*, Freiburg: University of Freiburg.
Pittock, Jamie (2011), 'A Pale Reflection of Political Reality: Integration of Global Climate, Wetland, and Biodiversity Agreements', *Climate Law*, 1(3), 343–373.

PointCarbon (2006), 'Senate funds AP6 after all' (3 July 2006), Oslo: PointCarbon.

Pollack, Mark A. and Gregory M. Shaffer (2012), 'The Interaction of Formal and Informal Lawmaking', in Joost Pauwelyn, Ramses A. Wessel and Jan Wouters (eds.), *Informal International Lawmaking*, Oxford: Oxford University Press, 241–270.

Pontecorvo, Concetta Maria (1999), 'Interdependence between Global Environmental Regimes: The Kyoto Protocol on Climate Change and Forest Protection', *Zeitschrift für ausländisches öffentliches Recht und Völkerrecht*, 59(3), 709–749.

Prins, Gwyn and Steve Rayner (2007), 'Time to Ditch Kyoto', *Nature*, 449(25), 973–975.

Prost, Mario (2008), 'All Shouting the Same Slogans: International Law's Unities and the Politics of Fragmentation', *Finnish Yearbook of International Law*, 17, 131–159.

Prost, Mario (2012), *The Concept of Unity in Public International Law*, Oxford: Hart.

Quick, Reinhard (2008), '"Border Tax Adjustment" in the Context of Emission Trading: Climate Protection or "Naked" Protectionism', *Global Trade and Customs Journal*, 3(5), 163–175.

Rahel, Frank J. and Julian D. Olden (2008), 'Assessing the Effects of Climate Change on Aquatic Invasive Species', *Conservation Biology*, 22(3), 521–533.

Rajamani, Lavanya (2006), *Differential Treatment in International Environmental Law*, Oxford: Oxford University Press.

Rajamani, Lavanya (2010), 'The Making and Unmaking of the Copenhagen Accord', *International and Comparative Law Quarterly*, 59(3), 824–843.

Rajamani, Lavanya (2011), 'The Cancun Climate Change Agreements: Reading the Text, Subtext and Tealeaves', *International and Comparative Law Quarterly*, 60(2), 499–519.

Rajamani, Lavanya (2012), 'The Durban Platform on Enhanced Action and the Future of the Climate Regime', *International and Comparative Law Quarterly*, 61(2), 501–518.

Raustiala, Kal (2005), 'Form and Substance in International Agreements', *American Journal of International Law*, 99(3), 581–614.

Raustiala, Kal (2013), 'Institutional Proliferation and the International Legal Order', in Jeffrey L. Dunoff and Mark A. Pollack (eds.), *Interdisciplinary Perspectives on International Law and International Relations: The State of the Art*, Cambridge, UK: Cambridge University Press, 293–320.

Raustiala, Kal and David G. Victor (1996), 'Biodiversity since Rio: The Future of the Convention on Biological Diversity', *Environment*, 38(4), 16–45.

Raustiala, Kal and David G. Victor (2004), 'The Regime Complex for Plant Genetic Resources', *International Organization*, 58(2), 277–309.

Rayner, Jeremy, Alexander Buck and Pia Katila (eds.) (2010), *Embracing Complexity: Meeting the Challenges of International Forest Governance. A Global Assessment Report Prepared by the Global Forest Expert Panel on the International Forest Regime*, Vienna: International Union of Forest Research Organizations.

Rechsteiner, Stefan, Christa Pfister and Fabian Martens (2009), 'TRIMS and the Clean Development Mechanism – Potential Conflicts', in Thomas Cottier, Olga Nartova and Sadeq Z. Bigdeli (eds.), *International Trade Regulation and the Mitigation of Climate Change*, Cambridge, UK: Cambridge University Press, 298–318.

Reinaud, Julia (2008), *Issues Behind Competitiveness and Carbon Leakage: Focus on Heavy Industry*, Paris: OECD/International Energy Agency.

Rittel, Horst W.J. and Melvin M. Webber (1973), 'Dilemmas in a General Theory of Planning', *Policy Sciences*, 4(2), 155–169.

Röben, Volker (2000), 'Institutional Developments under Modern International Environmental Agreements', *Max Planck Yearbook of United Nations Law*, 4, 363–443.

Rogelj, Joeri, Julia Nabel, Claudine Chen, William Hare, Kathleen Markmann, Malte Meinshausen, Michiel Schaeffer, Kirsten Macey and Nikas Höhne (2010), 'Copenhagen Pledges are Paltry', *Nature*, 464(7292), 1126–1128.

Rogelj, Joeri, William Hare, Jason Lowe, Detlef P. van Vuuren, Keywan Riahi, Ben Matthews, Tatsuya Hanaoka, Kejun Jiang and Malte Meinshausen (2011), 'Emission Pathways Consistent with a 2°C Global Temperature Limit', *Nature Climate Change*, 1, 413–418.

Rosenau, James N. (1995), 'Governance in the Twenty-first Century', *Global Governance*, 1(1), 13–43.

Rosenau, James and Ernst-Otto Czempiel (1992), *Governance without Government: Order and Change in World Politics*, Cambridge, UK: Cambridge University Press.

Rosendal, G. Kristin (2001), 'Impacts of Overlapping International Regime: The Case of Biodiversity', *Global Governance*, 7(1), 95–117.

Rosenthal, Erika and Robert Watson (2011), 'Multilateral Efforts to Reduce Black Carbon Emissions: A Lifeline for the Warming Arctic?', *Review of European Community and International Environmental Law*, 20(1), 3–10.

Ruis, Barbara (2001), 'No Forest Convention but Ten Tree Treaties', *Unasylva*, 52(3), 3–13.

Sadat-Akhavi, Seyed Ali (2003), *Methods of Resolving Conflicts between Treaties*, Leiden: Martinus Nijhoff.

Safrin, Sabrina (2002), 'Treaties in Collision? The Biosafety Protocol and the World Trade Organization Agreements', *American Journal of International Law*, 96(3), 606–628.

Sagemüller, Imke (2006), 'Forest Sinks under the United Nations Framework Convention on Climate Change and the Kyoto Protocol: Opportunity or Risk for Biodiversity?', *Columbia Journal of Environmental Law*, 31(2), 189–242.

Samson, Mélanie (2011), 'High Hopes, Scant Resources: A Word of Scepticism about the Anti-Fragmentation Function of Article 31(3)(c) of the Vienna Convention on the Law of Treaties', *Leiden Journal of International Law*, 24(3), 701–714.

Sands, Philippe (1999), 'Treaty, Custom and the Cross-fertilization of International Law', *Yale Human Rights and Development Law Journal*, 1(1), 85–105.

Savaresi, Annalisa (2011), 'Reducing Emissions from Deforestation in Developing Countries under the UNFCCC: Caveats and Opportunities for Biodiversity', *Yearbook of International Environmental Law*, 21, 81–113.

Schachter, Oscar (1977), 'The Twilight Existence of Nonbinding International Agreements', *American Journal of International Law*, 71(2), 296–304.

Schneider, Malte, Andreas Holzer and Volker Hoffmann (2008), 'Understanding the CDM's Contribution to Technology Transfer', *Energy Policy*, 36(8), 2930–2938.

Schoenbaum, Thomas J. (1997), 'International Trade and Protection of the Environment: The Continuing Search for Reconciliation', *American Journal of International Law*, 91(2), 268–313.

Schwartz, Jason (2006), '"Whose Woods These Are I Think I Know": How Kyoto May Change Who Controls Biodiversity', *New York University Environmental Law Journal*, 14(2), 421–480.

Scott, Joanne and Lavanya Rajamani (2012), 'EU Climate Change Unilateralism', *European Journal of International Law*, 23(2), 469–494.

Scott, Karen N. (2005), 'The Day after Tomorrow: Ocean CO_2 Sequestration and the Future of Climate Change', *Georgetown International Environmental Law Review*, 18(1), 57–108.

Scott, Karen N. (2011), 'International Environmental Governance: Managing Fragmentation through Institutional Connection', *Melbourne Journal of International Law*, 12(1), 177–216.

Secretariat of the CBD (2003), *Interlinkages between Biological Diversity and Climate Change. Advice on the Integration of Biodiversity Considerations into the Implementation of the United Nations Framework Convention on Climate Change and Its Kyoto Protocol*, Montreal: Secretariat of the Convention on Biological Diversity.

Secretariat of the CBD (2009a), *Review of the Literature on the Links Between Biodiversity and Climate Change: Impacts, Adaptation and Mitigation*, CBD Technical Series No. 42, Montreal: Secretariat of the CBD.

Secretariat of the CBD (2009b), *Connecting Biodiversity and Climate Change Mitigation and Adaptation. Report of the Second Ad Hoc Technical Expert Group on Biodiversity and Climate Change*, CBD Technical Series No. 41, Montreal: Secretariat of the CBD.

Selin, Henrik and Stacy D. VanDeveer (2003), 'Mapping Institutional Linkages in European Air Pollution Politics', *Global Environmental Politics*, 3(3), 14–46.

Seres, Stephen, Erik Haites and Kevin Murphy (2009), 'Analysis of Technology Transfer in CDM Projects: An Update', *Energy Policy*, 37(11), 4919–4926.

Shaffer, Gregory C. and Mark A. Pollack (2010), 'Hard Vs. Soft Law: Alternatives, Complements, and Antagonists in International Governance', *Minnesota Law Review*, 94(3), 706–799.

Shaffer, Gregory C. and Mark A. Pollack (2011), 'Hard Versus Soft Law in International Security', *Boston College Law Review*, 52(4), 1147–1241.

Shaffer, Gregory and Mark A. Pollack (2013), 'Hard and Soft Law: What Have We Learned?', in Jeffrey L. Dunoff and Mark A. Pollack (eds.), *Interdisciplinary Perspectives on International Law and International Relations: The State of the Art*, Cambridge, UK: Cambridge University Press, 197–222.

Shellenberger, Michael, Ted Nordhaus, Jeff Navin, Teryn Norris and Aden Van Noppen (2008), 'Fast, Clean, & Cheap: Cutting Global Warming's Gordian Knot', *Harvard Law & Policy Review*, 2, 93–118.

Shelton, Dinah (2006), 'Normative Hierarchy in International Law', *American Journal of International Law*, 100(2), 291–323.

Shelton, Dinah (2011), 'Comments on the Normative Challenge of Environmental "Soft Law"', in Yann Kerbrat and Sandrine Maljean-Dubois (eds.), *The Transformation of International Environmental Law*, Oxford: Hart, 61–73.

Simma, Bruno (1985), 'Self-Contained Regimes', *Netherlands Yearbook of International Law*, 16, 111–136.

Simma, Bruno (2004), 'Fragmentation in a Positive Light', *Michigan Journal of International Law*, 25(4), 845–847.

Simma, Bruno and Dirk Pulkowski (2006), 'Of Planets and the Universe: Self-contained Regimes in International Law', *European Journal of International Law*, 17(3), 483–529.

Sindico, Francesco (2009), 'Climate and Trade in a Divided World: Can Measures Adopted in the North End Up Shaping Climate Change Legislative Frameworks in the South?', in Benjamin J. Richardson, Yves Le Bouthillier, Heather McLeod-Kilmurray and Stepan Wood (eds.), *Climate Law and Developing Countries: Legal and Policy Challenges for the World Economy*, Cheltenham, UK and Northmapton, MA: Edward Elgar, 361–385.

Singh Ghaleigh, Navraj (2011), 'Barriers to Climate Technology Transfer – The Chimera of Intellectual Property Rights', *Carbon and Climate Law Review*, 4(2), 220–235.

Skjærseth, Jon Birger (2006), 'Protecting the Northeast Atlantic: One Problem, Three Institutions', in Sebastian Oberthür and Thomas Gehring (eds.), *Institutional Interaction in Global Environmental Governance. Synergy and Conflict among International and EU Policies*, Cambridge, MA: The MIT Press, 103–126.

Skjærseth, Jon Birger, Olav Schram Stokke and Jørgen Wettestad (2006), 'Soft Law, Hard Law, and Effective Implementation of International Environmental Norms', *Global Environmental Politics*, 6(3), 104–120.

Slaughter, Anne-Marie, Andrew S. Tulumello and Stepan Wood (1998), 'International Law and International Relations Theory: A New Generation of Interdisciplinary Scholarship', *American Journal of International Law*, 92(3), 367–397.

Slaughter-Burley, Anne-Marie (1993), 'International Law and International Relations Theory: A Dual Agenda', *American Journal of International Law*, 87(2), 205–239.

Sprinz, Detlef and Martin Weiß (2001), 'Domestic Politics and Global Climate Policy', in Urs Luterbacher and Detlef Sprinz (eds.), *International Relations and Global Climate Change*, Cambridge, MA: The MIT Press, 67–94.

Stern, Nicholas (2006), *The Stern Review: The Economics of Climate Change*, Cambridge, UK: Cambridge University Press.

Stiglitz, Joseph (2006), 'A New Agenda for Global Warming', *The Economists' Voice*, 3(7), 3.

Stockwell, Claire, William Hare and Kirsten Macey (2009), 'Designing a REDD Mechanism: The TDERM Triptych', in Benjamin J. Richardson, Yves Le Bouthillier, Heather McLeod-Kilmurray and Stepan Wood (eds.), *Climate Law and Developing Countries: Legal and Policy Challenges for the World Economy*, Cheltenham, UK and Northampton, MA: Edward Elgar, 151–177.

Stokke, Olav Schram (2001), *The Interplay of International Regimes: Putting Effectiveness Theory to Work?*, FNI Report 10/2001, Lysaker: The Fridtjof Nansen Institute.

Stokke, Olav Schram (2004), 'Trade Measures and Climate Compliance: Institutional Interplay Between WTO and the Marrakesh Accords', *International Environmental Agreements: Politics, Law and Economics*, 4(4), 339–357.

Stokke, Olav Schram and Sebastian Oberthür (2011), 'Introduction: Institutional Interaction in Global Environmental Change', in Sebastian Oberthür and Olav Schram Stokke (eds.), *Managing Institutional Complexity: Regime Interplay and Global Environmental Change*, Cambridge, MA: The MIT Press, 1–23.

Tamiotti, Ludivine (2011), 'The Legal Interface between Carbon Border Measures and Trade Rules', *Climate Policy*, 11(5), 1202–1211.

Tamura, Kentaro (2006), 'Technology Development and Transfer', in Ancha Srinivasan (ed.), *Asian Aspirations for Climate Regime Beyond 2012*, Hayama: Institute for Global Environmental Strategies, 53–76.

Tawney, Letha and Lutz Weischer (2011), *Innovation and Technology Transfer: Supporting Low Carbon Development with Climate Finance*, Washington, DC: World Resources Institute.

Teubner, Gunther (2012), *Constitutional Fragments: Societal Constitutionalism and Globalization*, Oxford: Oxford University Press.

The Economist (2013), 'The World Trade Organisation: Doha Delivers', The Economist (9 December 2013), available at http://www.economist.com/blogs/freeexchange/2013/12/world-trade-organisation (accessed 11 December 2013).

Third World Network (2009), 'G77/China Criticize Developed Countries' Trade Measures as Being against Convention', TWN Bonn News Update (14 August 2009).

Thomas, Sebastian, Paul Dargusch, Steve Harrison and John Herbohn (2010), 'Why Are There So Few Afforestation-based CDM Projects?', *Land Use Policy*, 27(3), 880–887.

Thompson, Dick (2001), 'Why U.S. Environmentalists Pin Hopes on Europe', *Time* (26 March 2001), available at http://www.time.com/time/world/article/0,8599,103985,00.html (accessed 11 December 2013).

Tiberghien, Yves and Miranda A. Schreurs (2007), 'High Noon in Japan: Embedded Symbolism and Post-2001 Kyoto Protocol Politics', *Global Environmental Politics*, 7(4), 70–91.

Trachtman, Joel P. (1999), 'The Domain of WTO Dispute Resolution', *Harvard International Law Journal*, 40(2), 333–377.

Trachtman, Joel P. (2002), 'Institutional Linkage: Transcending "Trade and …"', *American Journal of International Law*, 96(1), 77–93.

Trachtman, Joel P. (2011), 'Fragmentation and Coherence in International Law', available at http://papers.ssrn.com/sol3/papers.cfm?abstract_id=1908862 (accessed 11 December 2013).

Ulfstein, Geir (2007), 'Treaty Bodies', in Daniel Bodansky, Jutta Brunnée and Ellen Hey (eds.), *The Oxford Handbook of International Environmental Law*, Oxford: Oxford University Press, 877–889.

UNEP (1997), *Nairobi Declaration on the Role and Mandate of the United Nations Environment Programme*, UN Doc. UNEP/GC.19/1 (7 February 1997).

UNEP (2001), *Implementing the Clustering Strategy for Multilateral Environmental Agreements: A Framework. Background Paper by the Secretariat*, UN Doc. UNEP/IGM/4/4 (16 November 2001).

UNEP (2010), *Report of the Simultaneous Extraordinary Meetings of the Conferences of the Parties to the Basel, Rotterdam and Stockholm Conventions*, UN Doc. UNEP/FAO/CHW/RC/POPS/EXCOPS.1/8 (7 April 2010).

UNEP (2013), *The Emissions Gap Report 2013*, Nairobi: UNEP.

UNEP and International Institute for Sustainable Development (2005), *Environment and Trade: A Handbook*, 2nd Edition, Geneva/Winnipeg: UNEP and International Institute for Sustainable Development.

UNEP and WTO (2009), *Trade and Climate Change. A Report by the United Nations Environment Programme and the World Trade Organization*, Geneva: UNEP/WTO.

UNFCCC (2000), *Report of the Subsidiary Body on Scientific and Technological Advice on the Second Part of Its Thirteenth Session, The Hague, 13–18 November 2000*, UN Doc. FCCC/SBSTA/2000/14 (20 December 2000).

UNFCCC (2001a), *Cooperation with Relevant International Organizations. Submissions from Parties*, UN Doc. FCCC/SBSTA/2001/MISC.8/Add.2 (2 November 2001).

UNFCCC (2001b), *Report of the Subsidiary Body for Scientific and Technological Advice on its Fourteenth Session, Bonn, 24–27 July 2001*, UN Doc. FCCC/SBSTA/2001/2 (18 September 2001).

UNFCCC (2002), *Cooperation with Relevant International Organizations. Submissions from Parties*, UN Doc. FCCC/SBSTA/2002/MISC.9 (9 April 2002).

UNFCCC (2003), *Cooperation with Relevant International Organizations: World Trade Organization, Note by the Secretariat*, UN Doc. FCCC/SBSTA/2003/INF.7 (14 May 2003).

UNFCCC (2004), *Report of the Fifth Meeting of the Joint Liaison Group*, UN Doc. FCCC/SBSTA/2004/INF.9 (15 June 2004).

UNFCCC (2006a), *Report of the Conference of the Parties on Its Eleventh Session, Held at Montreal from 28 November to 10 December 2005*, UN Doc. FCCC/CP/2005/5 (30 March 2006).

UNFCCC (2006b), *Views on the Paper on Options for Enhanced Cooperation among the Three Rio Conventions, Submissions from Parties*, UN Doc. FCCC/SBSTA/2006/MISC.4 (23 March 2006).

UNFCCC (2007), *Innovative Options for Financing the Development and Transfer of Technologies*, Bonn: UNFCCC.

UNFCCC (2008), *Ideas and proposals on the elements contained in paragraph 1 of the Bali Action Plan. Submissions from Parties*, UN Doc. FCCC/AWGLCA/2008/MISC.2 (14 August 2008).

UNFCCC (2009), *Recommendations on Future Financing Options for Enhancing the Development, Deployment, Diffusion and Transfer of Technologies under the Convention. Report by the Chair of the Expert Group on Technology Transfer*, UN Doc. FCCC/SB/2009/2 (26 May 2009).

UNFCCC (2010a), *Report on Options to Facilitate Collaborative Technology Research and Development. Note by the Chair of the Expert Group on Technology Transfer*, UN Doc. FCCC/SBSTA/2010/INF.11 (24 November 2010).

UNFCCC (2010b), *Report on the Review and Assessment of the Effectiveness of the Implementation of Article 4, Paragraphs 1(c) and 5, of the Convention. Note by the Secretariat*, UN Doc. FCCC/SBI/2010/INF.4 (26 May 2010).

UNFCCC (2010c), *Negotiation Text. Note by the Secretariat*, UN Doc. FCCC/AWGLCA/2010/14 (13 August 2010).

UNFCCC (2011a), *Expert Workshop on the Technology Mechanism in Conjunction with the Fourteenth Session of the Ad Hoc Working Group on Long-term Cooperative Action under the Convention. Report by the Chair of the Workshop*, UN Doc. FCCC/AWGLCA/2011/INF.2 (26 May 2011).

UNFCCC (2011b), *Proposals by India for Inclusion of Additional Agenda Items in the Provisional Agenda of the Seventeenth Session of the Conference of the Parties. Note by the Secretariat*, UN Doc. FCCC/CP/2011/INF.2/Add. (7 October 2011).

UNFCCC (2013), *Compilation of Information on Mitigation Benefits of Actions, Initiatives and Options to Enhance Mitigation Ambition*, UN Doc. FCCC/TP/2013/4 (28 May 2013).

UNGA (1992a), *Agenda 21*, UN Doc. A/CONF.151/26/Rev.1 (14 June 1992).

UNGA (1992b), *Non-Legally Binding Authoritative Statement of Principles for a Global Consensus on the Management, Conservation and Sustainable Development of All Types of Forests* (13 June 1992), 31 *ILM* 881.
UNGA (1999), *United Nations General Assembly Resolution 53/186, International Institutional Arrangements Related to Environment and Development*, UN Doc. A/RES/53/18 (2 February 1999).
UNGA (2012), *United Nations General Assembly Resolution 66/288, The Future We Want*, UN Doc. A/RES/66/288 (11 September 2012).
United Nations Development Programme (2010), *Handbook for Conducting Technology Needs Assessment for Climate Change*, New York, NY: United Nations Development Programme.
van Aaken, Anne (2008), 'Fragmentation of International Law: The Case of International Investment Law', *Finnish Yearbook of International Law*, 17, 91–130.
van Asselt, Harro (2007), 'From UN-ity to Diversity? The UNFCCC, the Asia-Pacific Partnership, and the Future of International Law on Climate Change', *Carbon and Climate Law Review*, 1(1), 17–28.
van Asselt, Harro (2009), 'The EU ETS in the European Climate Policy Mix: Past, Present and Future', available at http://papers.ssrn.com/sol3/papers.cfm?abstract_id=1596892 (accessed 11 December 2013).
van Asselt, Harro (2010), 'Emissions Trading: The Enthusiastic Adoption of an Alien Instrument?', in Andrew Jordan, Dave Huitema, Harro van Asselt, Tim Rayner and Frans Berkhout (eds.), *Climate Change Policy in the European Union: Confronting the Dilemmas of Mitigation and Adaptation?*, Cambridge, UK: Cambridge University Press, 125–144.
van Asselt, Harro (2011a), 'Integrating Biodiversity in the Climate Regime's Forest Rules: Options and Tradeoffs in Greening REDD Design', *Review of European Community and International Environmental Law*, 20(2), 139–149.
van Asselt, Harro (2011b). 'Legal and Political Approaches in Interplay Management: Dealing with the Fragmentation of Global Climate Governance', in Sebastian Oberthür and Olav Schram Stokke (eds.), *Managing Institutional Complexity: Regime Interplay and Global Environmental Change*, Cambridge, MA: The MIT Press, 59–85.
van Asselt, Harro (2012a), 'Managing the Fragmentation of International Climate Law', in Erkki Hollo, Kati Kulovesi and Michael Mehling (eds.), *Climate Change and the Law*, Berlin: Springer, 329–357.
van Asselt, Harro (2012b), 'Managing the Fragmentation of International Environmental Law: Forests at the Intersection of the Climate and Biodiversity Regimes', *New York University Journal of International Law and Politics*, 44(4), 1205–1278.

van Asselt, Harro and Frank Biermann (2007), 'European Emissions Trading and the International Competitiveness of Energy-intensive Industries: A Legal and Political Evaluation of Possible Supporting Measures', *Energy Policy*, 35(1), 497–506.

van Asselt, Harro and Thomas L. Brewer (2010), 'Addressing Competitiveness and Leakage Concerns in Climate Policy: An Analysis of Border Adjustment Measures in the US and the EU', *Energy Policy*, 38(1), 42–51.

van Asselt, Harro and Joyeeta Gupta (2009), 'Stretching Too Far: Developing Countries and the Role of Flexibility Mechanisms beyond Kyoto', *Stanford Environmental Law Journal*, 28(2), 311–378.

van Asselt, Harro and Fariborz Zelli (2014, forthcoming). 'Connect the Dots: Managing the Fragmentation of Global Climate Governance', *Environmental Economics and Policy Studies* (in press).

van Asselt, Harro, Joyeeta Gupta and Frank Biermann (2005), 'Advancing the Climate Agenda: Exploiting Material and Institutional Linkages to Develop a Menu of Policy Options', *Review of European Community and International Environmental Law*, 14(3), 255–264.

van Asselt, Harro, Nicolien van der Grijp and Frans Oosterhuis (2006), 'Greener Public Purchasing: Opportunities for Climate-Friendly Government Procurement under WTO and EU Rules', *Climate Policy*, 6(2), 217–229.

van Asselt, Harro, Francesco Sindico and Michael A. Mehling (2008), 'Global Climate Change and the Fragmentation of International Law', *Law & Policy*, 30(4), 423–449.

van Asselt, Harro, Norichika Kanie and Masahiko Iguchi (2009), 'Japan's Position in International Climate Policy: Navigating between Kyoto and the APP', *International Environmental Agreements: Politics, Law and Economics*, 9(3), 319–336.

Van Damme, Isabelle (2009), *Treaty Interpretation by the WTO Appellate Body*, Oxford: Oxford University Press.

van der Gaast, Wytze (2005), 'Could the Asia-Pacific Partnership Bridge the EU–USA/Australia Gap?', *Joint Implementation Quarterly*, August 2005, 3–4.

van der Werf, G.R., D.C. Morton, R.S. DeFries, J.G.J. Olivier, P.S. Kasibhatla, R.B. Jackson, G.J. Collatz and J.T. Randerson (2009), 'CO_2 Emissions from Forest Loss', *Nature Geoscience*, 2, 737–738.

Velasquez, Jerry and Uli Piest (2003), 'Case Studies on Inter-linkages and Environmental Governance in 14 Asia and Pacific Countries', *Global Environmental Change*, 13(1), 61–68.

Velders, Guus J.M., Stephen O. Andersen, John S. Daniel, David W. Fahey and Mack McFarland (2007), 'The Importance of the Montreal

Protocol in Protecting Climate', *Proceedings of the National Academy of Sciences*, 104(12), 4814–4819.

Venter, Oscar, William F. Laurance, Takuya Iwamura, Kerrie A. Wilson, Richard A. Fuller and Hugh P. Possingham (2009), 'Harnessing Carbon Payments to Protect Biodiversity', *Science*, 326(5958), 1368.

Verchot, Louis V. and Elena Petkova (2009), *The State of REDD Negotiations: Consensus Points, Options for Moving Forward and Research Needs to Support the Process*, Bogor: Center for International Forestry Research.

Verheyen, Roda (2005), *Climate Change Damage and International Law: Prevention Duties and State Responsibility*, Leiden: Martinus Nijhoff.

Victor, David G. (2011), *Global Warming Gridlock: Creating More Effective Strategies for Protecting the Planet*, Cambridge, UK: Cambridge University Press.

Victor, David G., Kal Raustiala and Eugene B. Skolnikoff (eds.) (1998), *The Implementation and Effectiveness of International Environmental Commitments: Theory and Practice*, Cambridge, MA: The MIT Press.

Vierdag, E.W. (1988), 'The Time of the "Conclusion" of a Multilateral Treaty: Article 30 of the Vienna Convention on the Law of Treaties and Related Provisions', *British Yearbook of International Law*, 59, 75–111.

Vihma, Antto (2009), 'Friendly Neighbor or Trojan Horse? Assessing the Interaction of Soft Law Initiatives and the UN Climate Regime', *International Environmental Agreements: Politics, Law and Economics*, 9(3), 239–262.

Visseren-Hamakers, Ingrid J., Bas Arts and Pieter Glasbergen (2011), 'Interaction Management by Partnerships: The Case of Biodiversity and Climate Change', *Global Environmental Politics*, 11(3), 89–107.

Voigt, Christina (2008a), *Sustainable Development as a Principle of International Law. Resolving Conflicts between Climate Measures and WTO Law*, Leiden: Brill.

Voigt, Christina (2008b), 'WTO Law and International Emissions Trading: Is There Potential for Conflict?', *Carbon and Climate Law Review*, 2(1), 52–64.

von Bogdandy, Armin, Philipp Dann and Matthias Goldmann (2010), 'Developing the Publicness of Public International Law: Towards a Legal Framework for Global Governance Activities', in Armin von Bogdandy, Rüdiger Wolfrum, Jochen von Bernstorff, Philipp Dann and Matthias Goldmann (eds.), *The Exercise of Public Authority by International Institutions: Advancing International Institutional Law*, Berlin: Springer, 3–32.

von Moltke, Konrad (2005), 'Clustering International Environmental Agreements as an Alternative to a World Environment Organization', in

Frank Biermann and Steffen Bauer (eds.), *A World Environment Organization: Solution or Threat for Effective International Environmental Governance?*, Aldershot, UK: Ashgate, 175–204.

von Stein, Jana (2008), 'The International Law and Politics of Climate Change: Ratification of the United Nations Framework Convention and the Kyoto Protocol', *Journal of Conflict Resolution*, 52(2), 243–268.

Vranes, Erich (2009), *Trade and the Environment. Fundamental Issues in International and WTO Law*, Oxford: Oxford University Press.

Vranes, Erich (2011), 'Climate Labelling and the WTO: The 2010 EU Ecolabelling Programme as a Test Case Under WTO Law', *European Yearbook of International Economic Law*, 2(1), 205–223.

Weber, Christopher L. and Glen P. Peters (2009), 'Climate Change Policy and International Trade Policy Considerations in the US', *Energy Policy*, 37(2), 432–440.

Weiler, Joseph H.H. and Andreas Paulus (1997), 'The Structure of Change in International Law or Is There a Hierarchy of Norms in International Law?', *European Journal of International Law*, 8(4), 545–565.

Werksman, Jacob (1999a), 'Formal Linkages and Multilateral Environmental Agreements', available at http://archive.unu.edu/inter-linkages/1999/docs/jake.PDF (accessed 11 December 2013).

Werksman, Jacob (1999b), 'Greenhouse Gas Emissions Trading and the WTO', *Review of European Community and International Environmental Law*, 8(3), 251–264.

Werksman, Jacob (2001), 'Greenhouse-Gas Emissions Trading and the WTO', in W. Bradnee Chambers (ed.), *Inter-linkages. The Kyoto Protocol and the International Trade and Investment Regimes*, Tokyo: United Nations University Press, 153–190.

Werksman, Jacob (2009a), 'Balancing Investors' Interests and Global Policy Objectives in a Carbon Constrained World: The Interface of International Economic Law with the Clean Development Mechanism', in Thomas Cottier, Olga Nartova and Sadeq Z. Bigdeli (eds.), *International Trade Regulation and the Mitigation of Climate Change*, Cambridge, UK: Cambridge University Press, 319–327.

Werksman, Jacob (2009b), 'How Should a Post-2012 Climate Agreement Address Trade-Related Environmental Measures?', in Benjamin Simmons, Harro van Asselt and Fariborz Zelli (eds.), *Climate and Trade Policies in a Post-2012 World*, Geneva: UNEP, 27–33.

Werksman, Jacob and Trevor Houser (2008), *Competitiveness, Leakage and Comparability: Disciplining The Use of Trade Measures Under a Post-2012 Climate Agreement*, Washington, DC: World Resources Institute.

Werksman, Jacob, Kevin A. Baumert and Navroz K. Dubash (2003), 'Will International Investment Rules Obstruct Climate Protection Policies?', *International Environmental Agreements: Politics, Law and Economics*, 3(1), 59–86.

Wessel, Ramses (2011), 'Informal International Law-Making as a New Form of World Legislation?', *International Organizations Law Review*, 8(1), 253–265.

Wessel, Ramses, Joost Pauwelyn and Jan Wouters (2011), 'The Exercise of Public Authority through Informal International Law-Making', paper prepared for the conference Global Governance as Public Authority: Structures, Contestation, and Normative Change, Hertie School of Governance, Berlin, 15–16 April 2011.

White House (2001), 'President Bush Discusses Climate Change', available at http://georgewbush-whitehouse.archives.gov/news/releases/2001/06/20010611-2.html (accessed 11 December 2013).

Wiers, Jochem (2008), 'French Ideas on Climate and Trade Policies', *Carbon and Climate Law Review*, 2(1), 18–32.

Wiersema, Annecoos (2009), 'The New International Law-makers? Conferences of the Parties to Multilateral Environmental Agreements', *Michigan Journal of International Law*, 31(1), 231–287.

Williams, Angela (2008), 'Turning the Tide: Recognizing Climate Change Refugees in International Law', *Law & Policy*, 30(4), 504–529.

Wolfers, Arnold (1951), 'The Pole of Power and the Pole of Indifference', *World Politics*, 4(1), 39–63.

Wolfrum, Rüdiger and Nele Matz (2003), *Conflicts in International Environmental Law*, Berlin: Springer.

World Bank (2008a), *Development and Climate Change: A Strategic Framework for the World Bank Group*, Washington, DC: World Bank.

World Bank (2008b), *Governance Framework for the Strategic Climate Fund*, Washington, DC: World Bank.

World Resources Institute (2011), Climate Analysis Indicators Tool (CAIT) Version 6.0, Washington, DC: World Resources Institute.

WTO (1994), *Ministerial Decision on Trade and Environment*, WTO Doc. MTN.TNC/45(MIN), Annex (6 May 1994).

WTO (2001), *Doha Ministerial Declaration*, WTO Doc. WT/MIN(01)/DEC/1 (20 November 2001).

WTO (2007), *Existing Forms of Cooperation and Information Exchange between UNEP/MEAs and the WTO. Note by the Secretariat*, WTO Doc. TN/TE/S/2/Rev.2 (16 January 2007).

WTO (2008), *Chair's Summary – Informal Trade Ministers Dialogue on Climate Change Issues, Jimbaran, Bali, 8–9 December 2007. Communication from Indonesia*, WTO Doc. WT/L/717 (25 January 2008).

WTO (2009a), *Committee on Trade and Environment, Report of the Meeting Held on 3 November 2008. Note by the Secretariat*, WTO Doc. WT/CTE/M/46 (12 January 2009).

WTO (2009b), *Committee on Trade and Environment, Report of the Meeting Held on 10 July 2009. Note by the Secretariat*, WTO Doc. WT/CTE/M/47 (31 August 2009).

WTO (2011a), *Matrix on Trade Measures Pursuant to Selected Multilateral Environmental Agreements. Note by the Secretariat*, WTO Doc. WT/CTE/W/160/Rev.5; TN/TE/S/5/Rev.3 (15 June 2011).

WTO (2011b), *Report (2011) of the Committee on Trade and Environment*, WTO Doc. WT/CTE/18 (21 November 2011).

WTO (2011c), *Committee on Trade and Environment in Special Session. Report by the Chairman, Ambassador Manuel A.J. Teehankee, to the Trade Negotiations Committee*, WTO Doc. TN/TE/20 (21 April 2011).

Yamin, Farhana and Joanna Depledge (2004), *The International Climate Change Regime: A Guide to Rules, Institutions and Procedures*, Cambridge, UK: Cambridge University Press.

Yang, Chi-Yen, and Michael Oppenheimer (2007), 'A "Manhattan Project" for Climate Change', *Climatic Change*, 80(3–4), 199–204.

Yearwood, Ronnie R.F. (2011), *The Interaction between World Trade Organisation (WTO) Law and External International Law: The Constrained Openness of WTO Law (A Prologue to a Theory)*, Abingdon, UK: Routledge.

Young, Margaret A. (2007), 'The WTO's Use of Relevant Rules of International Law: An Analysis of the Biotech Case', *International and Comparative Law Quarterly*, 56(4), 907–930.

Young, Margaret A. (2011a), 'Climate Change Law and Regime Interaction', *Carbon and Climate Law Review*, 4(2), 147–157.

Young, Margaret A. (2011b), *Trading Fish, Saving Fish: The Interaction between Regimes in International Law*, Cambridge, UK: Cambridge University Press.

Young, Margaret A. (2012a), 'Introduction: The Productive Friction Between Regimes', in Margaret A. Young (ed.), *Regime Interaction in International Law: Facing Fragmentation*, Cambridge, UK: Cambridge University Press, 1–19.

Young, Margaret A. (ed.) (2012b), *Regime Interaction in International Law: Facing Fragmentation*, Cambridge, UK: Cambridge University Press.

Young, Oran R. (1996), 'Institutional Linkages in International Society: Polar Perspectives', *Global Governance*, 2(1), 1–24.

Young, Oran R. (1999a), *Governance in World Affairs*, Ithaca, NY: Cornell University Press.

Young, Oran R. (ed.) (1999b), *The Effectiveness of International Environmental Regimes: Causal Connections and Behavioral Mechanisms*, Cambridge, MA: The MIT Press.

Young, Oran R. (2002), *The Institutional Dimension of Environmental Change: Fit, Interplay and Scale*, Cambridge, MA: The MIT Press.

Young, Oran R. (2008), 'Deriving Insights from the Case of the WTO and the Cartagena Protocol', in Oran R. Young, W. Bradnee Chambers, Joy A. Kim and Claudia ten Have (eds.), *Institutional Interplay: Biosafety and Trade*, Tokyo: United Nations University Press, 131–158.

Young, Oran R. (2011), 'Effectiveness of International Environmental Regimes: Existing Knowledge, Cutting-edge Themes, and Research Strategies', *Proceedings of the National Academy of Sciences*, 108(50), 19853–19860.

Young, Oran R., Frans Berkhout, Gilberto C. Gallopin, Marco A. Janssen, Elinor Ostrom and Sander van der Leeuw (2006), 'The Globalization of Socio-ecological Systems: An Agenda for Scientific Research', *Global Environmental Change*, 16(3), 304–316.

Young, Oran R., W. Bradnee Chambers, Joy A. Kim and Claudia ten Have (eds.) (2008), *Institutional Interplay: Biosafety and Trade*. Tokyo: United Nations University Press.

Yu, Vicente Paolo B. (2009), *Developing Country Perspectives on Carbon-Based Competitiveness, Trade and Climate Change Linkages*, London, UK: Royal Institute of International Affairs.

Zane, Steven Nathaniel (2011), 'Leveling the Playing Field: The International Legality of Carbon Tariffs in the EU', *Boston College International and Comparative Law Review*, 34(1), 199–225.

Zelli, Fariborz (2009), *The Regime Environment of Environmental Regimes: Conceptualizing, Theorizing and Examining Conflicts among International Regimes on Environmental Issues*, PhD thesis, University of Tübingen.

Zelli, Fariborz (2011a), 'The Fragmentation of the Global Climate Governance Architecture', *WIREs Climate Change*, 2(2), 255–270.

Zelli, Fariborz (2011b), 'Regime Conflicts and their Management in Global Environmental Governance', in Sebastian Oberthür and Olav Schram Stokke (eds.), *Managing Institutional Complexity: Regime Interplay and Global Environmental Change*, Cambridge, MA: The MIT Press, 199–226.

Zelli, Fariborz and Harro van Asselt (2010), 'The Overlap between the UN Climate Regime and the World Trade Organization: Lessons for Post-2012 Climate Governance', in Frank Biermann Philipp Pattberg and Fariborz Zelli (eds.), *Global Climate Governance Beyond 2012: Architecture, Agency and Adaptation*, Cambridge, UK: Cambridge University Press, 79–96.

Zelli, Fariborz and Harro van Asselt (2013), 'The Institutional Fragmentation of Global Environmental Governance – Causes, Consequences and Responses', *Global Environmental Politics*, 13(3), 1–13.

Zelli, Fariborz, Aarti Gupta and Harro van Asselt (2012), 'Horizontal Institutional Interlinkages', in Frank Biermann and Philipp Pattberg (eds.), *Global Environmental Governance Reconsidered*, Cambridge, MA: The MIT Press, 175–198.

Zhang, ZhongXiang (2009), 'Encouraging Developing Country Involvement in a Post-2012 Climate Change Regime: Carrots, Sticks or Both?', in Benjamin Simmons, Harro van Asselt and Fariborz Zelli (eds.), *Climate and Trade Policies in a Post-2012 World*, Geneva: UNEP, 79–86.

Index